Selected Titles in This Series

700 **Vicente Cortés,** A new construction of homogeneous quaternionic manifolds and related geometric structures, 2000

699 **Alexander Fel′shtyn,** Dynamical zeta functions, Nielsen theory and Reidemeister torsion, 2000

698 **Andrew R. Kustin,** Complexes associated to two vectors and a rectangular matrix, 2000

697 **Deguang Han and David R. Larson,** Frames, bases and group representations, 2000

696 **Donald J. Estep, Mats G. Larson, and Roy D. Williams,** Estimating the error of numerical solutions of systems of reaction-diffusion equations, 2000

695 **Vitaly Bergelson and Randall McCutcheon,** An ergodic IP polynomial Szemerédi theorem, 2000

694 **Alberto Bressan, Graziano Crasta, and Benedetto Piccoli,** Well-posedness of the Cauchy problem for $n \times n$ systems of conservation laws, 2000

693 **Doug Pickrell,** Invariant measures for unitary groups associated to Kac-Moody Lie algebras, 2000

692 **Mara D. Neusel,** Inverse invariant theory and Steenrod operations, 2000

691 **Bruce Hughes and Stratos Prassidis,** Control and relaxation over the circle, 2000

690 **Robert Rumely, Chi Fong Lau, and Robert Varley,** Existence of the sectional capacity, 2000

689 **M. A. Dickmann and F. Miraglia,** Special groups: Boolean-theoretic methods in the theory of quadratic forms, 2000

688 **Piotr Hajłasz and Pekka Koskela,** Sobolev met Poincaré, 2000

687 **Guy David and Stephen Semmes,** Uniform rectifiability and quasiminimizing sets of arbitrary codimension, 2000

686 **L. Gaunce Lewis, Jr.,** Splitting theorems for certain equivariant spectra, 2000

685 **Jean-Luc Joly, Guy Metivier, and Jeffrey Rauch,** Caustics for dissipative semilinear oscillations, 2000

684 **Harvey I. Blau, Bangteng Xu, Z. Arad, E. Fisman, V. Miloslavsky, and M. Muzychuk,** Homogeneous integral table algebras of degree three: A trilogy, 2000

683 **Serge Bouc,** Non-additive exact functors and tensor induction for Mackey functors, 2000

682 **Martin Majewski,** ational homotopical models and uniqueness, 2000

681 **David P. Blecher, Paul S. Muhly, and Vern I. Paulsen,** Categories of operator modules (Morita equivalence and projective modules, 2000

680 **Joachim Zacharias,** Continuous tensor products and Arveson's spectral C^*-algebras, 2000

679 **Y. A. Abramovich and A. K. Kitover,** Inverses of disjointness preserving operators, 2000

678 **Wilhelm Stannat,** The theory of generalized Dirichlet forms and its applications in analysis and stochastics, 1999

677 **Volodymyr V. Lyubashenko,** Squared Hopf algebras, 1999

676 **S. Strelitz,** Asymptotics for solutions of linear differential equations having turning points with applications, 1999

675 **Michael B. Marcus and Jay Rosen,** Renormalized self-intersection local times and Wick power chaos processes, 1999

674 **R. Lawther and D. M. Testerman,** A_1 subgroups of exceptional algebraic groups, 1999

673 **John Lott,** Diffeomorphisms and noncommutative analytic torsion, 1999

672 **Yael Karshon,** Periodic Hamiltonian flows on four dimensional manifolds, 1999

671 **Andrzej Rosłanowski and Saharon Shelah,** Norms on possibilities I: Forcing with trees and creatures, 1999

670 **Steve Jackson,** A computation of δ^1_5, 1999

(*Continued in the back of this publication*)

A New Construction of Homogeneous Quaternionic Manifolds and Related Geometric Structures

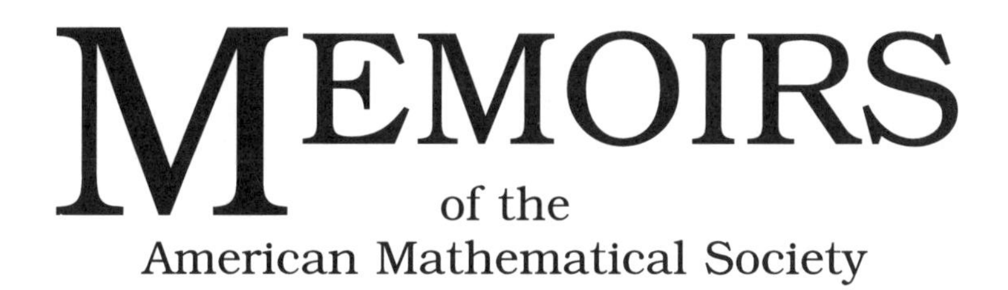

Number 700

A New Construction of Homogeneous Quaternionic Manifolds and Related Geometric Structures

Vicente Cortés

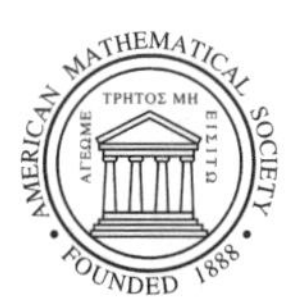

September 2000 • Volume 147 • Number 700 (end of volume) • ISSN 0065-9266

American Mathematical Society
Providence, Rhode Island

2000 *Mathematics Subject Classification.*
Primary 53C30; Secondary 53C25.

Library of Congress Cataloging-in-Publication Data

Cortés, Vicente, 1965–
A new construction of homogeneous quaternionic manifolds and related geometric structures / Vicente Cortés.
p. cm. — (Memoirs of the American Mathematical Society, ISSN 0065-9266 ; no. 700)
"Volume 147, number 700 (end of volume)."
Includes bibliographical references.
ISBN 0-8218-2111-3 (alk. paper)
1. Kèhlerian manifolds. I. Title. II. Series.
QA3 .A57 no. 700
[QA649]
510 s—dc21
[516.3′62] 00-034993

Memoirs of the American Mathematical Society

This journal is devoted entirely to research in pure and applied mathematics.

Subscription information. The 2000 subscription begins with volume 143 and consists of six mailings, each containing one or more numbers. Subscription prices for 2000 are $466 list, $419 institutional member. A late charge of 10% of the subscription price will be imposed on orders received from nonmembers after January 1 of the subscription year. Subscribers outside the United States and India must pay a postage surcharge of $30; subscribers in India must pay a postage surcharge of $43. Expedited delivery to destinations in North America $35; elsewhere $130. Each number may be ordered separately; *please specify number* when ordering an individual number. For prices and titles of recently released numbers, see the New Publications sections of the *Notices of the American Mathematical Society.*

Back number information. For back issues see the *AMS Catalog of Publications.*

Subscriptions and orders should be addressed to the American Mathematical Society, P. O. Box 845904, Boston, MA 02284-5904. *All orders must be accompanied by payment.* Other correspondence should be addressed to Box 6248, Providence, RI 02940-6248.

Memoirs of the American Mathematical Society is published bimonthly (each volume consisting usually of more than one number) by the American Mathematical Society at 201 Charles Street, Providence, RI 02904-2294. Periodicals postage paid at Providence, RI. Postmaster: Send address changes to Memoirs, American Mathematical Society, P. O. Box 6248, Providence, RI 02940-6248.

This publication is indexed in *Science Citation Index®*, *SciSearch®*, *Research Alert®*, *CompuMath Citation Index®*, *Current Contents®/Physical, Chemical & Earth Sciences.*
Printed in the United States of America.

♾ The paper used in this book is acid-free and falls within the guidelines established to ensure permanence and durability.
Visit the AMS home page at URL: `http://www.ams.org/`

10 9 8 7 6 5 4 3 2 1 05 04 03 02 01 00

Contents

Abstract viii
Introduction 1
1. Extended Poincaré algebras 7
2. The homogeneous quaternionic manifold (M, Q) associated to an extended Poincaré algebra 17
3. Bundles associated to the quaternionic manifold (M, Q) 34
4. Homogeneous quaternionic supermanifolds associated to superextended Poincaré algebras 39
Appendix. Supergeometry 44
Bibliography 59

Abstract

Let $V = \mathbb{R}^{p,q}$ be the pseudo-Euclidean vector space of signature (p,q), $p \geq 3$ and W a module over the even Clifford algebra $C\ell^0(V)$. A homogeneous quaternionic manifold (M,Q) is constructed for any $\mathfrak{spin}(V)$-equivariant linear map $\Pi : \wedge^2 W \to V$. If the skew symmetric vector valued bilinear form Π is nondegenerate then (M,Q) is endowed with a canonical pseudo-Riemannian metric g such that (M,Q,g) is a homogeneous quaternionic pseudo-Kähler manifold. If the metric g is positive definite, i.e. a Riemannian metric, then the quaternionic Kähler manifold (M,Q,g) is shown to admit a simply transitive solvable group of automorphisms. In this special case ($p = 3$) we recover all the known homogeneous quaternionic Kähler manifolds of negative scalar curvature (Alekseevsky spaces) in a unified and direct way. If $p > 3$ then M does not admit any transitive action of a solvable Lie group and we obtain new families of quaternionic pseudo-Kähler manifolds. Then it is shown that for $q = 0$ the noncompact quaternionic manifold (M,Q) can be endowed with a Riemannian metric h such that (M,Q,h) is a homogeneous quaternionic Hermitian manifold, which does not admit any transitive solvable group of isometries if $p > 3$.

The twistor bundle $Z \to M$ and the canonical $\mathrm{SO}(3)$-principal bundle $S \to M$ associated to the quaternionic manifold (M,Q) are shown to be homogeneous under the automorphism group of the base. More specifically, the twistor space is a homogeneous complex manifold carrying an invariant holomorphic distribution $\mathcal{D}$ of complex codimension one, which is a complex contact structure if and only if Π is nondegenerate. Moreover, an equivariant open holomorphic immersion $Z \to \bar{Z}$ into a homogeneous complex manifold $\bar{Z}$ of complex algebraic group is constructed.

Finally, the construction is shown to have a natural mirror in the category of supermanifolds. In fact, for any $\mathfrak{spin}(V)$-equivariant linear map $\Pi : \vee^2 W \to V$ a homogeneous quaternionic supermanifold (M,Q) is constructed and, moreover, a homogeneous quaternionic pseudo-Kähler supermanifold (M,Q,g) if the symmetric vector valued bilinear form Π is nondegenerate.

Key words: Quaternionic Kähler manifolds, twistor spaces, complex contact manifolds, homogeneous spaces, supermanifolds

1991 *Mathematics Subject Classification.* Primary: 53C30, Secondary: 53C25

Introduction

Let us start this introduction by recalling the notion of quaternionic manifold, see [**A-M2**]. A hypercomplex structure on a real vector space E consists of 3 complex structures (J_1, J_2, J_3) on E satisfying $J_1 J_2 = J_3$. It defines on E the structure of (left-) vector space over the quaternions $\mathbb{H} = \{1, i, j, k\}$ such that multiplication by i, j and k is given, respectively, by J_1, J_2 and J_3. The 3-dimensional subspace $Q = \mathrm{span}\{J_1, J_2, J_3\} \subset \mathrm{End}(E)$ is what is called a quaternionic structure on E. A Euclidean scalar product $\langle\cdot,\cdot\rangle$ on (E, Q) is called (Q-) Hermitian if Q consists of skew symmetric endomorphisms of $(E, \langle\cdot,\cdot\rangle)$. Now let M be a smooth manifold, $\dim M > 4$. An almost quaternionic structure Q on M is a smooth field $m \mapsto Q_m$ whose value at $m \in M$ is a quaternionic structure on $T_m M$. Q is called a quaternionic structure and (M, Q) a **quaternionic manifold** if there exists a torsionfree connection on TM preserving the rank 3 subbundle $Q \subset \mathrm{End}(TM)$. Now let g be a Riemannian metric on M, Hermitian with respect to an (almost) quaternionic structure Q. Then M with the structure (Q, g) is called an (almost) quaternionic Hermitian manifold. If, moreover, the Levi-Civita connection preserves Q then (M, Q, g) is said to be a **quaternionic Kähler manifold**. We remark that quaternionic Kähler manifolds represent one of the few basic Riemannian geometries, as defined by Berger's list of possible Riemannian holonomy groups, see [**A1**], [**Bes**], [**Br1**] and [**S2**]. For the possible holonomy groups of, not necessarily Riemannian, torsionfree connections see [**Br2**], [**Schw**] and references therein.

Next we review what is known about homogeneous quaternionic Kähler manifolds. First of all, quaternionic Kähler manifolds (M, Q, g) are Einstein manifolds, i.e. $Ric = cg$, see e.g. [**A1**], [**Bes**], [**S2**] and [**Kak**]. We discuss 3 cases depending on the sign of the constant c (which is the sign of the scalar curvature).

$c = 0$) Simply connected Ricci-flat quaternionic Kähler manifolds are hyper-Kähler manifolds, see [**Bes**] Ch. 14. These are Kähler manifolds with respect to 3 complex structures J_1, J_2 and $J_3 = J_1 J_2$. It is a general fact that any homogeneous Ricci-flat Riemannian manifold is necessarily flat, see [**A-K**]. In particular, all homogeneous hyper-Kähler manifolds are flat. For hyper-Kähler manifolds of small cohomogeneity see [**Bi1**], [**Bi2**], [**B-G**], [**D-S1**], [**D-S2**], [**K-S2**], [**Sw1**], [**Sw2**] and [**C4**].

$c > 0$) It follows from Myer's theorem that any complete Einstein manifold of positive scalar curvature is necessarily compact. In particular, any complete quaternionic Kähler manifold of positive scalar curvature is compact. It was proven in [**L-S**] that for every $n > 1$ there is only a finite number of such manifolds of dimension $4n$ up to homothety, cf. [**Bea**], [**G-S**], [**L1**], [**L4**], [**L5**], [**P-S**]. The only known examples are, up to now, the Wolf spaces [**W1**]. These are precisely the homogeneous quaternionic Kähler manifolds of positive scalar curvature and are all symmetric of compact type, cf. [**A2**]. More generally, the Wolf spaces can be characterized as the compact quaternionic Kähler manifolds which admit an action of cohomogeneity ≤ 1 by a compact semisimple group of isometries and which are not scalar-flat, see [**D-S3**], cf. [**A-P**].

$c < 0$) Complete noncompact quaternionic Kähler manifolds of negative scalar curvature exist in abundance. In fact, it was proven in [**L3**] that the moduli space of complete quaternionic Kähler metrics on $\mathbb{R}^{4n}$, $n > 1$, is infinite dimensional. For

Received by the editors March 1, 1999.

explicit constructions of, in general not complete, quaternionic Kähler manifolds see e.g. [**G1**], [**G-L**] and [**L2**].

What about homogeneous examples? First of all, the noncompact duals of the Wolf spaces are symmetric (and hence homogeneous) quaternionic Kähler manifolds of negative Ricci and nonpositive sectional curvature. Moreover, like any Riemannian symmetric space of noncompact type, these manifolds admit smooth quotients by discrete cocompact groups of isometries, see [**Bo**], [**Fi-S**]. The first examples of quaternionic Kähler manifolds which are not locally symmetric were found by D.V. Alekseevsky in [**A3**]. An Alekseevsky space is a homogeneous quaternionic Kähler manifold of negative scalar curvature which admits a simply transitive splittable solvable group of isometries. It follows from Iwasawa's decomposition theorem that the noncompact duals of the Wolf spaces are precisely the symmetric Alekseevsky spaces. Besides these there are 3 series of nonsymmetric Alekseevsky spaces, see [**A3**], [**dW-VP2**] and [**C2**]. In [**A3**] it was conjectured that any noncompact homogeneous quaternionic Kähler manifold admits a transitive solvable group of isometries. This conjecture is still open: up to now, the only known examples of homogeneous quaternionic Kähler manifolds of negative scalar curvature are the Alekseevsky spaces. However, by the construction presented in this work we obtain many homogeneous quaternionic pseudo-Kähler manifolds (with indefinite metric) which do not admit any transitive action of a solvable Lie group. Moreover, in 2.6 we construct a family of noncompact homogeneous quaternionic Hermitian manifolds (with positive definite metric) with no transitive solvable group of isometries.

It is natural to ask for examples of compact locally homogeneous quaternionic Kähler manifolds. The following negative result was proven in [**A-C4**]. Let M be a compact quaternionic Kähler manifold or, more generally, a quaternionic Kähler manifold of finite volume. If the universal cover $\tilde{M}$ is a homogeneous quaternionic Kähler manifold then it is necessarily symmetric. In particular, the only Alekseevsky spaces which admit smooth quotients of finite volume by discrete groups of isometries are the symmetric ones, this was as well proven in [**A-C1**] by a simpler method. Additionally, the symmetric Alekseevsky spaces can be characterized by the property of having nonpositive curvature, see [**C2**].

Given a simply transitive Lie group L of isometries acting on a Riemannian manifold M, there exists an algorithm to compute the full isometry group of M [**A-W**], cf. [**Wo**]. However, this algorithm involves the covariant derivatives (of all orders) of the curvature tensor and hence can only be applied effectively in very simple situations. If the simply transitive group L is splittable solvable and unimodular then the full isometry group is easily computed, see [**G-W**]. Unfortunately, the splittable solvable groups of isometries acting simply transitively on the Alekseevsky spaces are not unimodular. For that reason in [**A-C1**] a new algorithm was developed which completely avoids the curvature tensor and works also for nonunimodular splittable solvable groups. Using it the full isometry group of the nonsymmetric Alekseevsky spaces was determined [**A-C1**]. The Lie algebra of the full isometry group was previously described by the theoretical physicists de Wit, Vanderseypen and Van Proeyen by a different method, see [**dW-V-VP**].

In the following, we shortly comment on the attention paid to Alekseevsky's spaces in the physical literature. There is a concept of special geometry, which evolved in the theory of strings and supergravity, see e.g. [**Z**], [**B-W**], [**dW-VP1**], [**G-S-T**]. More specifically, special Kähler geometry is the geometry associated to $N = 2$ supergravity in $d = 4$ space time dimensions coupled to vector multiplets and was first described in [**dW-VP1**], cf. [**C-D-F**], [**St**], [**C3**] and [**Fr**]. The (Kuranishi) moduli space of a Calabi-Yau 3-fold bears this particular geometry. Moreover, there is a construction (the "c-map"), related to Mirror Symmetry, which to any special Kähler manifold associates a quaternionic Kähler manifold, see [**C-F-G**] and [**F-S**]; for the general framework see [**Cr**]. Using the c-map, Cecotti [**Ce**] was the first to relate the classification problem for homogeneous special Kähler manifolds to Alekseevsky's classification [**A3**]. In addition, he introduced Vinberg's theory of T-algebras [**V2**] to describe the first nonsymmetric homogeneous special Kähler manifolds (the symmetric special Kähler manifolds were described in terms of Jordan algebras, see [**C-VP**] and [**G-S-T**]). Cecotti's classification of homogeneous special Kähler manifolds was extended in [**dW-VP2**], [**C3**] and [**A-C5**]. In [**C4**], the hyper-Kählerian version of the c-map was used to construct a natural (pseudo-) hyper-Kähler structure on the bundle of intermediate Jacobians over the moduli space of gauged Calabi-Yau 3-folds.

In the last part of the introduction we describe the main components, results and the global structure of the present paper. The basic algebraic data of our construction are a pseudo-Euclidean vector space V, a module W over the even Clifford algebra $C\ell^0(V)$ and a $\mathfrak{spin}(V)$-equivariant linear map $\Pi : \wedge^2 W \to V$. Any such Π defines a $\mathbb{Z}_2$-graded Lie algebra $\mathfrak{p} = \mathfrak{p}(\Pi) = \mathfrak{p}_0 + \mathfrak{p}_1$, where $\mathfrak{p}_0 = \mathrm{Lie\,Isom}(V) = \mathfrak{o}(V) + V$ and $\mathfrak{p}_1 = W$. Here V acts trivially on W and $\mathfrak{o}(V) \cong \mathfrak{spin}(V)$ acts via the inclusion $\mathfrak{spin}(V) \subset C\ell^0(V)$ on the $C\ell^0(V)$-module W. The Lie bracket on $\mathfrak{p}_1 \times \mathfrak{p}_1$ is given by Π. The Lie algebras $\mathfrak{p}(\Pi)$ were introduced in [**A-C2**], where a basis for the vector space of $\mathfrak{spin}(V)$-equivariant linear maps $\Pi : \wedge^2 W \to V$ was explicitly constructed. The Lie algebra $\mathfrak{p}(\Pi)$ is called an extended Poincaré algebra of signature (p,q) if $V \cong \mathbb{R}^{p,q}$ has signature (p,q). A mirror symmetric version of this construction is obtained replacing $\Pi : \wedge^2 W \to V$ by a $\mathfrak{spin}(V)$-equivariant linear map $\Pi : \vee^2 W \to V$. Here $\vee^2 W = \mathrm{Sym}^2 W$ denotes the symmetric square of W. The corresponding algebraic structure $\mathfrak{p}(\Pi) = \mathfrak{p}_0 + \mathfrak{p}_1$ is now a super Lie algebra. It is called a superextended Poincaré algebra of signature (p,q). For special signatures (p,q) of space time V these super Lie algebras play an important role in the physical literature since the early days of supersymmetry and supergravity, see e.g. [**Go-L**] and [**O-S**]; for more recent contributions see e.g. [**F**] and references therein. Notice that if $(p,q) = (1,3)$ then $V = \mathbb{R}^{1,3}$ is Minkowski space and the even subalgebra $\mathfrak{p}_0 = \mathrm{Lie\,Isom}(\mathbb{R}^{1,3}) \subset \mathfrak{p}(\Pi)$ is the classical Poincaré algebra. The construction of all superextended Poincaré algebras (of arbitrary signature) was carried out in [**A-C2**]. In [**A-C-D-S**] the twistor equation is interpreted as the differential equation satisfied by infinitesimal automorphisms of a geometric structure modelled on the linear Lie supergroup associated to a superextended Poincaré algebra (for more on the twistor equation see e.g [**K-R**] and references therein). In contrast with the case of superextended Poincaré algebras, to our knowledge, extended Poincaré algebras do not occur in the physical literature before the publication of [**A-C2**]: The first occurrence is [**D-L**].

Any extended Poincaré algebra $\mathfrak{p}(\Pi)$, as above, admits a derivation D with eigenspace decomposition $\mathfrak{p}(\Pi) = \mathfrak{o}(V) + V + W$ and corresponding eigenvalues $(0, 1, 1/2)$. We can extend $\mathfrak{p}(\Pi)$ by D obtaining a new $\mathbb{Z}_2$-graded Lie algebra $\mathfrak{g} = \mathfrak{g}(\Pi) = \mathfrak{g}_0 + \mathfrak{g}_1$, where $\mathfrak{g}_0 = \mathbb{R}D + \mathfrak{p}_0 = \mathbb{R}D + \mathfrak{o}(V) + V$ and $\mathfrak{g}_1 = \mathfrak{p}_1 = W$. The adjoint representation of the Lie algebra $\mathfrak{g}$ is faithful and hence defines on $\mathfrak{g}$ the structure of linear Lie algebra. Let $G = G(\Pi) \subset \mathrm{Aut}\,\mathfrak{g}$ denote the corresponding connected linear Lie group. It is the numerator of the homogeneous quaternionic manifolds $M = M(\Pi) = G/K$ we are going to construct. To define the denominator K let $E \subset V$ be a 3-dimensional Euclidean subspace and $\mathfrak{k} = \mathfrak{k}(E) \subset \mathfrak{o}(V) \subset \mathfrak{g}$ the maximal subalgebra which preserves E, i.e. $\mathfrak{k} = \mathfrak{o}(E) \oplus \mathfrak{o}(E^{\perp})$. Now we define $K = K(E) \subset G$ to be the connected linear Lie group with Lie algebra $\mathfrak{k}$. Our main theorem is the following, see Thm. 8:

Theorem A *Let V be any pseudo-Euclidean vector space, $E \subset V$ a Euclidean 3-dimensional subspace, W any $C\ell^0(V)$-module and $\Pi : \wedge^2 W \to V$ any $\mathfrak{spin}(V)$-equivariant linear map. Let $M = G/K$ be the homogeneous manifold associated to the Lie groups $G = G(\Pi)$ and $K = K(E) \subset G$ constructed above. Then the following is true:*

1) *There exists a G-invariant quaternionic structure Q on M.*
2) *If Π is nondegenerate (i.e. if $W \ni s \mapsto \Pi(s \wedge \cdot) \in W^* \otimes V$ is injective) then there exists a G-invariant pseudo-Riemannian metric g on M such that (M, Q, g) is a homogeneous quaternionic pseudo-Kähler manifold.*

We remark that for $V = \mathbb{R}^{3,q}$ and W arbitrary the map Π can always be chosen such that the metric g in 2) becomes positive definite. In this special case we recover the 3 series of Alekseevsky spaces by a simple and unified construction, which completely avoids the technicalities of constructing complicated representations of Kählerian Lie algebras, see [**A3**] and [**C2**].

If in all constructions $\Pi : \wedge^2 W \to V$ is replaced by a $\mathfrak{spin}(V)$-equivariant linear map $\Pi : \vee^2 W \to V$ then we obtain the following analogue of Theorem A in the category of supermanifolds, see Thm. 17:

Theorem B *Let V be any pseudo-Euclidean vector space, $E \subset V$ a Euclidean 3-dimensional subspace, W any $C\ell^0(V)$-module and $\Pi : \vee^2 W \to V$ any $\mathfrak{spin}(V)$-equivariant linear map. Let $M = G/K$ be the homogeneous supermanifold associated to the Lie supergroups $G = G(\Pi)$ and $K = K(E) \subset G$ constructed in 4. Then the following holds:*

1) *There exists a G-invariant quaternionic structure Q on M.*
2) *If Π is nondegenerate (i.e. if $W \ni s \mapsto \Pi(s \vee \cdot) \in W^* \otimes V$ is injective) then there exists a G-invariant pseudo-Riemannian metric g on M such that (M, Q, g) is a homogeneous quaternionic pseudo-Kähler supermanifold.*

(For the definition of quaternionic structure, pseudo-Riemannian metric etc. on a supermanifold see the appendix.)

Furthermore, we remark that replacing Euclidean 3-space $E \cong \mathrm{R}^{3,0}$ by Lorentzian 3-space $E \cong \mathbb{R}^{1,2}$ one obtains a para-quaternionic version of our construction.

Finally, we outline the structure of the paper: In section 1 we discuss extended Poincaré algebras. The basic definitions and formulas are given in 1.1. To our fundamental map $\Pi : \wedge^2 W \to V$ and to an oriented Euclidean subspace $E \subset V$, $\dim E \equiv 3 \pmod 4$, we associate a canonical symmetric bilinear form b on W and study its properties in 1.2. Using the form b, in 1.3 we classify extended Poincaré algebras of signature (p, q), $p \equiv 3 \pmod 4$, up to isomorphism.

In section 2 we construct the homogeneous quaternionic manifolds of Theorem A. The basic notions of quaternionic geometry are recalled in 2.2. First, see 2.1, we describe the structure of the Lie group $G = G(\Pi)$ and the coset spaces $M =$

G/K, $K = K(E)$, where $E \subset V$ is any pseudo-Euclidean subspace. The proof of Theorem A is given in 2.4. A crucial observation is that $M = G/K$ contains the locally symmetric quaternionic pseudo-Kähler submanifold $M_0 = G_0/K$, where $G_0 \subset G$ is the connected linear Lie group associated to $\mathfrak{g}_0 \subset \mathfrak{g} = \mathfrak{g}_0 + \mathfrak{g}_1$. The first step consists in extending the G_0-invariant quaternionic and pseudo-Riemannian structures on M_0 to G-invariant structures on M. Using the canonical symmetric bilinear form b on W introduced in 1.2 the pseudo-Riemannian metric is extended, in the nondegenerate case, to a G-invariant pseudo-Riemannian metric g on M. The quaternionic structure is always extended by the beautifully simple formula (13) to a G-invariant almost quaternionic structure Q on M. To prove that Q is a quaternionic structure, we construct a G-invariant torsionfree connection ∇ on M which preserves Q. In the nondegenerate case ∇ is simply the Levi-Civita connection of the pseudo-Riemannian metric g. We use the description of invariant connections on homogeneous manifolds in terms of Nomizu maps, see 2.3.

Then we concentrate on the Riemannian case, see 2.5. The Riemannian manifolds $M(\Pi)$ are classified up to isometry using results of 1.3. We show that all these manifolds admit a non-Abelian simply transitive splittable solvable group of isometries and hence are Alekseevsky spaces, see Thm. 9. Moreover, we explain how to obtain the 3 series of Alekseevsky spaces by specifying V, W, and Π, see Thm. 10.

The quaternionic pseudo-Kähler manifolds $(M(\Pi), Q, g)$ of Theorem A do not admit any transitive action of a solvable group if $V = \mathbb{R}^{p,q}$ with $p > 3$, see Thm. 12. Moreover, if $q = 0$, then we can replace g by a G-invariant Riemannian and Q-Hermitian metric h such that $(M(\Pi), h, g)$ are homogeneous quaternionic Hermitian manifolds with no transitive solvable group of isometries, see Thm. 11.

In section 3 we study the various bundles associated to the quaternionic manifolds $M = M(\Pi) = G/K$: The twistor bundle $Z(M)$, the canonical $\mathrm{SO}(3)$-principal bundle $S(M)$ and the Swann bundle $U(M)$. We show that G acts transitively on $Z(M)$ and $S(M)$ and with cohomogeneity one on $U(M)$. In particular, $Z = Z(M)$ is a homogeneous complex manifold of the group $G = G(\Pi)$. We exhibit a G-invariant holomorphic tangent hyperplane distribution $\mathcal{D}$ on Z and prove that $\mathcal{D}$ defines a complex contact structure on Z if and only if Π is nondegenerate. Moreover, we construct an open G-equivariant holomorphic immersion $Z \to \bar{Z}$ of Z into a homogeneous complex manifold of the complex algebraic group $G^{\mathbb{C}} \subset \mathrm{Aut}(\mathfrak{g}^{\mathbb{C}})$, see Thm. 14. This immersion is a finite covering over an open G-orbit. In the nondegenerate case $\bar{Z}$ is a homogeneous complex contact manifold of the group $G^{\mathbb{C}}$ and the immersion is a morphism of complex contact manifolds.

In the final section 4 we extend our construction to the category of supermanifolds, proving Theorem B. We have aimed at a straightforward presentation, summarizing the needed supergeometric background in the appendix.

Acknowledgements It is a pleasure to thank my friend and coauthor Dmitry V. Alekseevsky for many intensive discussions during our productive collaboration. Also I am very grateful to Werner Ballmann and Ursula Hamenstädt for their encouragement and support. For reading the manuscript I thank Gregor Weingart. Finally, I am indebted to many collegues for hospitality and conversations, especially to Oliver Baues, Robert Bryant, Shiing-Shen Chern, Jost Eschenburg, Phillip A. Griffiths, Ernst Heintze, Alan Huckleberry, Enrique Macías, Robert Osserman, Hans-Bert Rademacher, Gudlaugur Thorbergsson and Joseph A. Wolf.

My research has received important support from the following institutions: SFB 256 (Bonn University), Alexander von Humboldt Foundation and Mathematical Sciences Research Institute (Berkeley).

1. Extended Poincaré algebras

1.1. Basic facts. Let V be a pseudo-Euclidean vector space with scalar product $\langle \cdot, \cdot \rangle$. There exists an orthonormal basis (e_i), $i = 1, \dots, n = \dim V = p + q$, of V such that $\langle x, x \rangle = \sum_{i=1}^{p} (x^i)^2 - \sum_{j=p+1}^{n} (x^j)^2$ for all $x = \sum_{i=1}^{n} x^i e_i \in V$. Any such basis defines an isometry between V and the standard pseudo-Euclidean vector space $\mathbb{R}^{p,q}$ of signature (p, q). The isometry group of V is the semidirect product

$$\mathrm{Isom}(V) = \mathrm{O}(V) \ltimes V .$$

DEFINITION 1. *The Lie group* $\mathrm{P}(V) := \mathrm{Isom}(V)$ *is called the* **Poincaré group** *of* V*. Its Lie algebra* $\mathfrak{p}(V) = \mathfrak{o}(V) + V$ *is called the* **Poincaré algebra** *of* V.

Next we recall some basic facts concerning the Clifford algebra $C\ell(V) = C\ell^0(V) + C\ell^1(V)$, see [**L-M**]. Any unit vector $x \in V$, $\langle x, x \rangle = \pm 1$, defines an invertible element $x \in C\ell(V)$. The group $\mathrm{Pin}(V) \subset C\ell(V)$ generated by all unit vectors is called the **pin group**. Its subgroup $\mathrm{Spin}(V) := \mathrm{Pin}(V) \cap C\ell^0(V)$ consisting of even elements is called the **spin group**. The **adjoint representation**

$$\mathrm{Ad} : \mathrm{Pin}(V) \longrightarrow \mathrm{O}(V) ,$$

$$\mathrm{Ad}(x) y = x y x^{-1} \in V , \quad x \in \mathrm{Pin}(V) , \quad y \in V ,$$

induces a two-fold covering of the special orthogonal group:

$$\mathrm{Spin}(V) \xrightarrow{2:1} \mathrm{SO}(V) .$$

In fact, if $r_x \in \mathrm{O}(V)$ denotes the reflection in the hyperplane $x^\perp \subset V$ orthogonal to a unit vector $x \in V$ then the following formula

$$\mathrm{Ad}(xy) = r_x \circ r_y \tag{1}$$

holds for any two unit vectors $x, y \in V$. The groups $\mathrm{Pin}(V)$ and $\mathrm{Spin}(V)$ are Lie groups whose Lie algebra is the Lie subalgebra $\mathfrak{spin}(V) \subset C\ell^0(V)$ generated by the commutators $[x, y] = xy - yx$ of all elements $x, y \in V$. It is canonically isomorphic to the orthogonal Lie algebra $\mathfrak{o}(V)$ via the adjoint representation

$$\mathrm{ad} = \mathrm{Ad}_* : \mathfrak{spin}(V) \xrightarrow{\sim} \mathfrak{o}(V) , \quad \mathrm{ad}(x) y = [x, y] , \quad x \in \mathfrak{spin}(V) , \quad y \in V . \tag{2}$$

In fact, if we identify $\wedge^2 V = \mathfrak{o}(V)$ by

$$(x \wedge y)(z) := \langle y, z \rangle x - \langle x, z \rangle y , \quad x, y, z \in V \tag{3}$$

then $\mathrm{ad}^{-1} : \wedge^2 V = \mathfrak{o}(V) \xrightarrow{\sim} \mathfrak{spin}(V)$ is given by the following equation:

$$\mathrm{ad}^{-1}(x \wedge y) = -\frac{1}{4}[x, y] , \quad x, y \in V . \tag{4}$$

In particular, $\mathrm{ad}(xy) = -2x \wedge y$ if x and y are orthogonal.

Any $C\ell(V)$-module W can be decomposed into irreducible submodules. Depending on the signature (p, q) of V, there exist one or two irreducible $C\ell(V)$-modules up to equivalence. In case there are two, they are related by the unique automorphism of $C\ell(V)$ which preserves V and acts on V as $-\mathrm{Id}$. The restriction of an irreducible $C\ell(V)$-module S to $C\ell^0(V)$ (respectively, to $\mathrm{Spin}(V)$ and $\mathfrak{spin}(V)$) is, up to equivalence, independent of W and is called the **spinor module** of $C\ell^0(V)$ (respectively, of $\mathrm{Spin}(V)$ and $\mathfrak{spin}(V)$). The spinor module S is either irreducible or $S = S^+ \oplus S^-$ is the sum of two irreducible **semispinor modules** $S^\pm$, which may be equivalent or not, depending on the signature of V. If S^+ and S^- are not equivalent, they are related by an automorphism of $C\ell^0(V)$ which preserves V and acts as an isometry on V. In the following we will freely use standard notations such as $C\ell_{p,q} = C\ell(\mathbb{R}^{p,q})$, $\mathrm{Spin}(p,q) = \mathrm{Spin}(\mathbb{R}^{p,q})$, $\mathrm{Spin}(p) = \mathrm{Spin}(p,0)$ etc., cf. [**L-M**].

Now let W be a module of the even Clifford algebra $C\ell^0(V)$ and $\Pi : \wedge^2 W \to V$ a $\mathfrak{spin}(V)$-equivariant linear map. Given these data we extend the Lie bracket on $\mathfrak{p}(V)$ to a Lie bracket $[\cdot,\cdot]$ on the vector space $\mathfrak{p}(V) + W$ by the following requirements:

1) W is a $\mathfrak{p}(V)$-submodule with trivial action of V and action of $\mathfrak{o}(V)$ defined by $\mathrm{ad}^{-1} : \mathfrak{o}(V) \to \mathfrak{spin}(V) \subset C\ell^0(V)$, see equation (2),
2) $[s,t] = \Pi(s \wedge t)$ for all $s, t \in W$.

The reader may observe that the Jacobi identity follows from 1) and 2). The resulting Lie algebra will be denoted by $\mathfrak{p}(\Pi)$. Note that $\mathfrak{p} = \mathfrak{p}(\Pi)$ has a $\mathbb{Z}_2$-grading

$$\mathfrak{p} = \mathfrak{p}_0 + \mathfrak{p}_1\,, \quad \mathfrak{p}_0 = \mathfrak{p}(V)\,, \quad \mathfrak{p}_1 = W\,,$$

compatible with the Lie bracket, i.e. $[\mathfrak{p}_a, \mathfrak{p}_b] \subset \mathfrak{p}_{a+b}$, $a, b \in \mathbb{Z}_2 = \mathbb{Z}/2\mathbb{Z}$. In other words, $\mathfrak{p}(\Pi)$ is a $\mathbb{Z}_2$-graded Lie algebra.

DEFINITION 2. *Any $\mathbb{Z}_2$-graded Lie algebra $\mathfrak{p}(\Pi) = \mathfrak{p}(V) + W$ as above is called an* **extended Poincaré algebra** *(of* **signature** (p,q) *if $V \cong \mathbb{R}^{p,q}$). $\mathfrak{p}(\Pi)$ is called* **nondegenerate** *if Π is nondegenerate, i.e. if the map $W \ni s \mapsto \Pi(s \wedge \cdot) \in W^* \otimes V$ is injective.*

The structure of extended Poincaré algebra on the vector space $\mathfrak{p}(V) + W$ is completely determined by the $\mathfrak{o}(V)$-equivariant map $\Pi : \wedge^2 W \to V$ ($\mathfrak{o}(V)$ acts on W via $\mathrm{ad}^{-1} : \mathfrak{o}(V) \to \mathfrak{spin}(V) \subset C\ell^0(V)$). The set of all $\mathfrak{o}(V)$-equivariant linear maps $\wedge^2 W \to V$ is naturally a vector space. In fact, it is the subspace $(\wedge^2 W^* \otimes V)^{\mathfrak{o}(V)}$ of $\mathfrak{o}(V)$-invariant elements of the vector space $\wedge^2 W^* \otimes V$ of all linear maps $\wedge^2 W \to V$. In the classification [**A-C2**] an explicit basis for the vector space $(\wedge^2 W^* \otimes V)^{\mathfrak{o}(V)}$ of extended Poincaré algebra structures on $\mathfrak{p}(V) + W$ is constructed for all possible signatures (p,q) of V and any $C\ell^0(V)$-module W.

1.2. The canonical symmetric bilinear form b. Let $V = \mathbb{R}^{p,q}$ be the standard pseudo-Euclidean vector space with scalar product $\langle \cdot,\cdot \rangle$ of signature (p,q). From now on we fix a decomposition $p = p' + p''$ and assume that $p' \equiv 3 \pmod 4$, see Remark 2 below.

Remark 1: Notice that p and q are on equal footing, since any extended Poincaré algebra of signature (q,p) is isomorphic to an extended Poincaré algebra of signature (p,q). In fact, the canonical antiisometry which maps the standard orthonormal basis of $\mathbb{R}^{q,p}$ to that of $\mathbb{R}^{p,q}$ induces an isomorphism of the corresponding Poincaré algebras which is trivially extended to an isomorphism of extended Poincaré algebras.

We denote by $(e_i) = (e_1, \ldots, e_{p'})$ the first p' basis vectors of the standard basis of V and by $(e_i') = (e_1', \ldots, e_{p''+q}')$ the remaining ones. The two complementary orthogonal subspaces of V spanned by these bases are denoted by $E = \mathbb{R}^{p'} = \mathbb{R}^{p',0}$ and $E' = E^\perp = \mathbb{R}^{p'',q}$ respectively. The vector spaces V, E and E' are oriented by their standard orthonormal bases. E.g, the orientation of Euclidean p'-space E defined by the basis (e_i) is $e_1^* \wedge \cdots \wedge e_{p'}^* \in \wedge^{p'} E^*$. Here (e_i^*) denotes the basis of E^* dual to (e_i). Now let $\mathfrak{p}(\Pi) = \mathfrak{p}(V) + W$ be an extended Poincaré algebra of signature (p,q) and $(\tilde{e}_i)$ any orhonormal basis of E. Then we define a $\mathbb{R}$-bilinear form $b_{\Pi,(\tilde{e}_i)}$ on the $\mathcal{C}\ell^0(V)$-module W by:

$$(5) \qquad b_{\Pi,(\tilde{e}_i)}(s,t) = \langle \tilde{e}_1, [\tilde{e}_2 \ldots \tilde{e}_{p'} s, t] \rangle = \langle \tilde{e}_1, \Pi(\tilde{e}_2 \ldots \tilde{e}_{p'} s \wedge t) \rangle \,, \quad s, t \in W \,.$$

We put $b = b(\Pi) := b_{\Pi,(e_i)}$ for the standard basis (e_i) of E.

Remark 2: Equation (5) defines a skew symmetric bilinear form on W if $p' \equiv 1 \pmod 4$. For even p' the above formula does not make sense, unless one assumes that W is a $\mathcal{C}\ell(V)$-module rather than a $\mathcal{C}\ell^0(V)$-module. Here we are only interested in the case $p' \equiv 3 \pmod 4$. Moreover, later on, for the construction of homogeneous quaternionic manifolds we will put $p' = 3$.

THEOREM 1. *The bilinear form b has the following properties:*

1) $b_{\Pi,(\tilde{e}_i)} = \pm b$ *if* $\tilde{e}_1 \wedge \cdots \wedge \tilde{e}_{p'} = \pm e_1 \wedge \cdots \wedge e_{p'}$. *In particular,* $b_{\Pi,(\tilde{e}_i)} = b$ *for any positively oriented orthonormal basis* $(\tilde{e}_i)$ *of* E.
2) b *is symmetric.*
3) b *is invariant under the maximal connected subgroup* $K(p',p'') = \mathrm{Spin}(p') \cdot \mathrm{Spin}_0(p'',q) \subset \mathrm{Spin}(p,q)$ *which preserves the orthogonal decomposition* $V = E + E'$ *(and is not* $\mathrm{Spin}_0(p,q)$*-invariant, unless* $p'' + q = 0$*).*
4) *Under the identification* $\mathfrak{o}(V) = \wedge^2 V = \wedge^2 E + \wedge^2 E' + E \wedge E'$*, see equation (3), the subspace* $E \wedge E'$ *acts on* W *by* b*-symmetric endomorphisms and the subalgebra* $\wedge^2 E \oplus \wedge^2 E' \cong \mathfrak{o}(p') \oplus \mathfrak{o}(p'',q)$ *acts on* W *by* b*-skew symmetric endomorphisms.*

Proof: Obviously 4) $\Rightarrow$ 3). We show first that 3) $\Rightarrow$ 1). If $(\tilde{e}_i)$ is a positively oriented orthonormal basis then there exists $\varphi \in \mathrm{Spin}(p')$ such that $\mathrm{Ad}(\varphi)e_i = \tilde{e}_i$, $i = 1, \ldots, p'$. Now $\Pi = [\cdot,\cdot]| \wedge^2 W : \wedge^2 W \to V$ is $\mathfrak{spin}(V)$-equivariant and hence equivariant under the connected group $\mathrm{Spin}_0(V) \subset \mathrm{Spin}(V)$. In particular, Π is $\mathrm{Spin}(p')$-equivariant. Under the condition 3), this implies that

$$\begin{aligned}
b(s,t) &= b(\varphi^{-1}s, \varphi^{-1}t) = \langle e_1, [e_2 \ldots e_{p'} \varphi^{-1} s, \varphi^{-1} t] \rangle \\
&= \langle e_1, [\varphi^{-1} \mathrm{Ad}_\varphi e_2 \ldots \mathrm{Ad}_\varphi e_{p'} s, \varphi^{-1} t] \rangle \\
&= \langle e_1, \mathrm{Ad}(\varphi^{-1}) [\tilde{e}_2 \ldots \tilde{e}_{p'} s, t] \rangle = \langle \tilde{e}_1, [\tilde{e}_2 \ldots \tilde{e}_{p'} s, t] \rangle \\
&= b_{\Pi,(\tilde{e}_i)}(s,t) \,, \quad s, t \in W \,.
\end{aligned}$$

Here we have used the notation $\mathrm{Ad}_\varphi = \mathrm{Ad}(\varphi)$. The case of negatively oriented orthonormal basis $(\tilde{e}_i)$ follows now from the Clifford relation $\tilde{e}_i \tilde{e}_j = -\tilde{e}_j \tilde{e}_i$, $i \neq j$.

Next we prove 2) using first the $\mathfrak{spin}(V)$-equivariance of Π then equation (4) and eventually $p' \equiv 3 \pmod 4$:

$$\begin{aligned}
b(t,s) &= \langle e_1, [e_2 \ldots e_{p'} t, s] \rangle \\
&= -\langle e_1, [e_4 \ldots e_{p'} t, e_2 e_3 s] \rangle + \langle e_1, \mathrm{ad}(e_2 e_3)[e_4 \ldots e_{p'} t, s] \rangle \\
&= -\langle e_1, [e_4 \ldots e_{p'} t, e_2 e_3 s] \rangle = \cdots = -\langle e_1, [t, e_2 \ldots e_{p'} s] \rangle \\
&= \langle e_1, [e_2 \ldots e_{p'} s, t] \rangle = b(s,t) \,.
\end{aligned}$$

Finally we prove 4). By equation (4) we have to check that $e_i e_j, e'_k e'_l \in \mathfrak{spin}(V)$ ($i \neq j$ and $k \neq l$) act by b-skew symmetric endomorphisms and $e_i e'_k$ by a b-symmetric endomorphism on W. This is done in the next computation, in which we use again the equivariance of Π and equations (1) and (4) to express the adjoint representations Ad and ad respectively:

$$\begin{aligned} b(e_1e_2s,t) &= \langle e_1, [e_2 \dots e_{p'}e_1e_2s, t]\rangle = \langle e_1, [e_1e_2e_2\dots e_{p'}s, e_1e_2e_1e_2t]\rangle \\ &= \langle e_1, \mathrm{Ad}(e_1e_2)[e_2\dots e_{p'}s, e_1e_2t]\rangle = -\langle e_1, [e_2\dots e_{p'}s, e_1e_2t]\rangle \\ &= -b(s, e_1e_2t)\,, \\ b(e_2e_3s,t) &= \langle e_1, [e_2\dots e_{p'}e_2e_3s, t]\rangle = \langle e_1, [e_2e_3e_2\dots e_{p'}s, t]\rangle \\ &= \langle e_1, \mathrm{ad}(e_2e_3)[e_2\dots e_{p'}s, t]\rangle - \langle e_1, [e_2\dots e_{p'}s, e_2e_3t]\rangle \\ &= -b(s, e_2e_3t)\,, \\ b(e'_ke'_ls,t) &= \langle e_1, [e_2\dots e_{p'}e'_ke'_ls, t]\rangle = \langle e_1, [e'_ke'_le_2\dots e_{p'}s, t]\rangle \\ &= -\langle e_1, [e_2\dots e_{p'}s, e'_ke'_lt]\rangle + \langle e_1, \mathrm{ad}(e'_ke'_l)[e_2\dots e_{p'}s, t]\rangle \\ &= -b(s, e'_ke'_lt) + 0 = -b(s, e'_ke'_lt)\,. \end{aligned}$$

This already proves 3) and hence 1); in particular we have:

$$b_{\Pi,(e_1,e_2,\dots,e_{p'})} = b_{\Pi,(e_2,\dots,e_{p'},e_1)}\,.$$

Due to this symmetry, it is sufficient to check that $e_2e'_k$ acts as b-symmetric endomorphism on W:

$$\begin{aligned} b(e_2e'_ks,t) &= \langle e_1, [e_2\dots e_{p'}e_2e'_ks, t]\rangle = \langle e_1, [e_3\dots e_{p'}e'_ks, t]\rangle \\ &\overset{(*)}{=} -\langle e_1, [s, e_3\dots e_{p'}e'_kt]\rangle = -\langle e_1, [s, e_2\dots e_{p'}e_2e'_kt]\rangle \\ &= \langle e_1, [e_2\dots e_{p'}e_2e'_kt, s]\rangle = b(e_2e'_kt, s) \\ &= b(s, e_2e'_kt)\,. \end{aligned}$$

At $(*)$ we have used $(p'-1)/2$ times the $\mathfrak{spin}(V)$-equivariance of Π and the fact that $(p'-1)/2$ is odd if $p' \equiv 3 \pmod 4$. □

DEFINITION 3. *The bilinear form* $b = b(\Pi) = b_{\Pi,(e_1,\dots,e_{p'})}$ *defined above is called the* **canonical symmetric bilinear form** *on* W *associated to the* $\mathfrak{spin}(V)$-*equivariant map* $\Pi : \wedge^2 W \to V = \mathbb{R}^{p,q}$ *and the decomposition* $p = p' + p''$.

PROPOSITION 1. *The kernels of the linear maps* $\Pi : W \to W^* \otimes V$ *and* $b = b(\Pi) : W \to W^*$ *coincide:* $\ker \Pi = \ker b$.

Proof: It follows from Thm. 1, 4) that $W_0 := \ker b \subset W$ is $\mathfrak{o}(V)$-invariant. This implies that $\Pi(W_0 \wedge W) \subset V$ is an $\mathfrak{o}(V)$-submodule. The definition of W_0 implies that $\Pi(W_0 \wedge W) \subset E' = E^\perp$ and hence by Schur's lemma $\Pi(W_0 \wedge W) = 0$. This proves that $\ker b \subset \ker \Pi$. On the other hand, we have the obvious inclusion $\ker \Pi \subset \ker(b \circ e_2 \dots e_{p'}) = \ker b$. Here the equation follows from the $\mathrm{Spin}(p')$-invariance of b, see Thm. 1, 3). □

COROLLARY 1. $\mathfrak{p}(\Pi)$ *is nondegenerate (see Def. 2) if and only if* $b(\Pi)$ *is nondegenerate.*

THEOREM 2. *Let* $\mathfrak{p}(\Pi) = \mathfrak{p}(V) + W$ *be any extended Poincaré algebra of signature* (p,q), $p = p' + p''$, $p' \equiv 3 \pmod 4$, *and* b *the canonical symmetric bilinear form associated to these data. Then there exists a* b-*orthogonal decomposition* $W = \oplus_{i=0}^{l+m} W_i$ *into* $C\ell^0(V)$-*submodules with the following properties*

1) $[W_0, W] = 0$, $[W_i, W_j] = 0$ *if* $i \neq j$ *and* $[W_i, W_i] = V$ *for all* $i = 1, 2, \dots, l$.
2) $W_0 = \ker b$ *and* W_i *is* b-*nondegenerate for all* $i \geq 1$.

3) *For* $i = 1, \dots, l$ *the* $\mathcal{C}\ell^0(V)$*-submodule* W_i *is irreducible and for* $j = l+1, \dots, l+m$ *the* $\mathcal{C}\ell^0(V)$*-submodule* $W_j = X_j \oplus X_j'$ *is the direct sum of two irreducible* b*-isotropic* $\mathcal{C}\ell^0(V)$*-submodules.*
4) *The restriction of* b *to a bilinear form on any irreducible* $\mathcal{C}\ell^0(V)$*-submodule of* $X = \oplus_{j=l+1}^{l+m} W_j$ *vanishes.*

Proof: By Prop. 1, $W_0 := \ker b = \ker \Pi$ satisfies $[W_0, W] = 0$. As kernel of the $\mathfrak{o}(V)$-equivariant map Π the subspace W_0 is $\mathfrak{o}(V)$-invariant and hence a $\mathcal{C}\ell^0(V)$-submodule. We denote by W' a complementary $\mathcal{C}\ell^0(V)$-submodule. Every such submodule is b-nondegenerate. Let $W_i \subset W'$ be any irreducible $\mathcal{C}\ell^0(V)$-submodule. By Thm. 1, 4), $\ker(b|W_i \times W_i)$ is $\mathfrak{o}(V)$-invariant and hence a $\mathcal{C}\ell^0(V)$-submodule. Now by Schur's lemma we conclude that either $b|W_i \times W_i = 0$ or b is nondegenerate on W_i. In particular, we can decompose $W' = \oplus_{i=1}^{l} W_i \oplus X$ as direct b-orthogonal sum of b-nondegenerate $\mathcal{C}\ell^0(V)$-submodules such that W_i is irreducible and the restriction of b to a bilinear form on any irreducible $\mathcal{C}\ell^0(V)$-submodule of X vanishes. Let $Y, Z \subset X$ be two such submodules, $Y \neq Z$. The bilinear form b induces a linear map $Y \to Z^*$. By Thm. 1, 4), the kernel of this map is $\mathfrak{o}(V)$-invariant and hence a $\mathcal{C}\ell^0(V)$-submodule. Now Schur's lemma implies that either the kernel is Y and hence the restriction of b to a bilinear form on $Y \oplus Z$ vanishes or the kernel is trivial and b is nondegenerate on $Y \oplus Z$. In the second case X splits as direct b-orthogonal sum: $X = (Y \oplus Z) \oplus (Y \oplus Z)^{\perp}$. This shows that $X = \oplus_{i=l+1}^{l+m} W_i$ is the direct orthogonal sum of b-nondegenerate $\mathcal{C}\ell^0(V)$-submodules W_i such that $W_i = X_i \oplus X_i'$ is the direct sum of two b-isotropic irreducible submodules. This proves 2), 3) and 4). Now 1) is established applying Schur's lemma to the $\mathfrak{o}(V)$-equivariant map Π. In fact, $b(W_i, W_j) = 0$ (respectively, $b(W_i, W_i) \neq 0$) implies $\Pi(W_i \wedge W_j) \subset E'$ (respectively, $\Pi(\wedge^2 W_i) \neq 0$) and thus $[W_i, W_j] = \Pi(W_i \wedge W_j) = 0$ (respectively, $[W_i, W_j] = \Pi(\wedge^2 W_i) = V$). □

Next we will construct the subgroup $\hat{K}(p', p'') \subset \mathrm{Spin}(p, q)$ which consists of all elements preserving the orthogonal decomposition $V = E + E'$ and the canonical symmetric bilinear form b on W. Its identity component is the group $K(p', p'') \subset \mathrm{Spin}_0(p, q)$ introduced above. We will see that if $p'' = 0$ then $\hat{K}(p', p'') = \hat{K}(p, 0)$ is a maximal compact subgroup of $\mathrm{Spin}(p, q)$, which together with the element $1 \in \mathcal{C}\ell^0_{p,q}$ generates the even Clifford algebra $\mathcal{C}\ell^0_{p,q}$. This property will be very useful in the next section.

We denote by $x \mapsto x'$ the linear map $\mathbb{R}^q = \mathbb{R}^{q,0} \to \mathbb{R}^{0,q}$ which maps the standard orthonormal basis $(e_1, \dots, e_q)$ of $\mathbb{R}^q$ to the standard orthonormal basis $(e_1', \dots, e_q')$ of $\mathbb{R}^{0,q}$. It is an antiisometry: $\langle x', x' \rangle = -\langle x, x \rangle$. Let $\omega_{p'} = e_1 \dots e_{p'}$ be the volume element of $\mathcal{C}\ell_{p'} = \mathcal{C}\ell_{p',0}$, $(e_1, \dots, e_{p'})$ the standard orthonormal basis of $\mathbb{R}^{p'} = \mathbb{R}^{p',0} \subset \mathbb{R}^{p,q}$. Note that, since p' is odd, the volume element $\omega_{p'}$ commutes with $\mathbb{R}^{p'}$ and anticommutes with $\mathbb{R}^{p'',q} \supset \mathbb{R}^{0,q}$. Moreover, it satisfies $\omega_{p'}^2 = 1$, due to $p' \equiv 3 \pmod 4$.

LEMMA 1. *The map*

$$\mathbb{R}^q \ni x \mapsto \omega_{p'} x' \in \mathcal{C}\ell^0_{p,q}$$

extends to an embedding $\iota : \mathcal{C}\ell_q \hookrightarrow \mathcal{C}\ell^0_{p,q}$ *of algebras, which restricts to an embedding* $\iota|\mathrm{Pin}(q) : \mathrm{Pin}(q) \hookrightarrow \mathrm{Spin}(p, q)$ *of groups.*

Proof: It follows from $(\omega_{p'} x')^2 = -\omega_{p'}^2 x'^2 = -x'^2 = \langle x', x' \rangle = -\langle x, x \rangle$ that the map $x \mapsto \omega_{p'} x'$ extends to a homomorphism ι of Clifford algebras. Recall that $\mathcal{C}\ell_q$ is either simple or the sum of two simple ideals. In the first case, we can immediately conclude that $\ker \iota$ is trivial and hence ι an embedding. In the second case, the two simple ideals of $C\ell_q$ are $C\ell_q^{\pm} := (1 \pm \omega_q)C\ell_q$, where $\omega_q = e_1 \dots e_q$ is the volume element of $C\ell_q$. Now it is sufficient to check that $\iota(1 \pm \omega_q) \neq 0$. Using the fact that q is odd if $C\ell_q$ is not simple we compute

$$\iota(\omega_q) = \pm\omega_{p'}^q e_1' e_2' \dots e_q' = \pm e_1 e_2 \dots e_{p'} e_1' e_2' \dots e_q' .$$

This shows that $\iota(1 \pm \omega_q) = 1 \pm \iota(\omega_q) \neq 0$. □

We denote by $\hat{K}(p', p'') \subset \mathrm{Spin}(p, q)$ the subgroup generated by the subgroups $\mathrm{Spin}(p') \cdot \mathrm{Spin}_0(p'', q) \subset \mathrm{Spin}(p, q)$ and $\iota(\mathrm{Pin}(q)) \subset \mathrm{Spin}(p, q)$.

THEOREM 3. *The group $\hat{K}(p', p'')$ has the following properties:*

1) *$\hat{K}(p', p'') \subset \mathrm{Spin}(p, q)$ consists of all elements preserving the orthogonal decomposition $V = E + E'$ and the canonical symmetric bilinear form b on W. Its identity component is the group $K(p', p'') = \mathrm{Spin}(p') \cdot \mathrm{Spin}_0(p'', q) = \hat{K}(p', p'') \cap \mathrm{Spin}_0(p, q)$.*
2) *The homogeneous space $\mathrm{Spin}(p, q)/\hat{K}(p', p'')$ is connected.*
3) *$\hat{K}(p', p'')$ is compact if and only if $q = 0$ or $p'' = 0$ (and hence $p = p'$). In the latter case $\hat{K}(p', p'') = \hat{K}(p, 0) = \mathrm{Spin}(p) \cdot \iota(\mathrm{Pin}(q)) \subset \mathrm{Spin}(p, q)$ is a maximal compact subgroup and $\hat{K}(p, 0) \cong (\mathrm{Spin}(p) \times \mathrm{Pin}(q))/\{\pm 1\}$. Finally, in this case, the even Clifford algebra $C\ell_{p,q}^0$ is generated by 1 and $\hat{K}(p, 0)$.*

Proof: The first part of 1) can be checked using Thm. 1, 4) and implies the second part of 1). To prove 2) it is sufficient to observe that $\iota(\mathrm{Pin}(q)) \cong \mathrm{Pin}(q)$ has nontrivial intersection with all connected components of $\mathrm{Spin}(p, q)$ (due to our assumption $p \geq 3$ there are two such components if $q \neq 0$). The first part of 3) now follows simply from the fact that $\mathrm{Spin}_0(p'', q)$ is compact if and only if $p'' = 0$ or $q = 0$. The compact group $\hat{K}(p, 0) \subset \mathrm{Spin}(p, q)$ is maximal compact, because it has the same number of connected components as $\mathrm{Spin}(p, q)$, and from $\mathrm{Spin}(p) \cap \iota(\mathrm{Pin}(q)) = \{\pm 1\}$ we obtain the isomorphism $\hat{K}(p, 0) = \mathrm{Spin}(p) \cdot \iota(\mathrm{Pin}(q)) \cong (\mathrm{Spin}(p) \times \mathrm{Pin}(q))/\{\pm 1\}$. Finally, to prove the last statement, one easily checks that $\hat{K}(p, 0)$ contains all quadratic monomials xy in unit vectors $x, y \in \mathbb{R}^{p,0} \cup \mathbb{R}^{0,q}$. □

COROLLARY 2. *The correspondence $\Pi \mapsto b(\Pi)$ defines an injective linear map $(\wedge^2 W^* \otimes V)^{\mathfrak{o}(V)} \to (\vee^2 W^*)^{\hat{K}(p', p'')}$.*

Proof: The existence of the map follows from Thm. 3. We prove the injectivity. From $b(\Pi) = 0$ it follows that $\Pi(\wedge^2 W) \subset E'$ and hence by Schur's lemma $\Pi = 0$. □

1.3. The set of isomorphism classes of extended Poincaré algebras. Starting from the decomposition proven in Thm. 2 we will derive the classification of extended Poincaré algebras of signature (p, q), $p \equiv 3 \pmod 4$, up to isomorphism. It will turn out that the space of isomorphism classes is naturally parametrized by a finite number of integers. We fix the decomposition $p = p' + p''$, $p' = p$, $p'' = 0$, and for any extended Poincaré algebra $\mathfrak{p}(\Pi)$ of signature (p, q) as above we consider the canonical symmetric bilinear form $b = b_{\Pi,(e_1, \dots e_p)}$.

THEOREM 4. *Let $\mathfrak{p}(\Pi) = \mathfrak{p}(V) + W$ be any extended Poincaré algebra of signature (p, q), $p \equiv 3 \pmod 4$. Then there exists a b-orthogonal decomposition $W = \oplus_{i=0}^l W_i$ into $C\ell^0(V)$-submodules with the following properties*

1) $[W_0, W] = 0$, $[W_i, W_j] = 0$ *if* $i \neq j$ *and* $[W_i, W_i] = V$ *for all* $i = 1, 2, \ldots, l$.
2) $W_0 = \ker b$ *and* W_i *is* b*-nondegenerate for all* $i \geq 1$.
3) W_i, $i \geq 1$, *is an irreducible* $\mathcal{C}\ell^0(V)$*-submodule on which* b *is (positive or negative) definite.*

Proof: Let $W = \oplus_{i=1}^{l+m} W_i$ be a decomposition as in Thm. 2. It only remains to prove that b is definite on W_i for $i = 1, \ldots, l$ and that $X = \oplus_{i=l+1}^{l+m} W_i = 0$. This follows from Lemma 3 and Lemma 4 below. □

LEMMA 2. *The restriction of an irreducible* $\mathcal{C}\ell^0_{p,q}$*-module* Σ *to a module of the maximal compact subgroup* $\hat{K} = \hat{K}(p, 0) = \mathrm{Spin}(p) \cdot \iota(\mathrm{Pin}(q)) \subset \mathrm{Spin}(p, q)$ *is irreducible. Here* $\iota : \mathrm{Pin}(p) \hookrightarrow \mathrm{Spin}(p, q)$ *is the embedding of Lemma 1. Moreover,* Σ *is irreducible as module of the connected group* $K = K(p, 0) = \mathrm{Spin}(p) \cdot \iota(\mathrm{Spin}(q)) = \mathrm{Spin}(p) \cdot \mathrm{Spin}(q)$ *if and only if* $n = p + q \equiv 2, 4, 5$ *or* $6 \pmod 8$. *If* $n \equiv 0, 1, 3$ *or* $7 \pmod 8$ *then* Σ *is the sum of two irreducible* K*-submodules.*

Proof: Recall that $\mathcal{C}\ell_{p,q} = \mathcal{C}\ell_p \hat{\otimes} \mathcal{C}\ell_{0,q}$ is (identified with) the $\mathbb{Z}_2$-graded tensor product of the Clifford algebras $\mathcal{C}\ell_p = \mathcal{C}\ell_{p,0}$ and $\mathcal{C}\ell_{0,q}$. It is easily checked, using the classification of Clifford algebras and their modules, see [**L-M**], that any irreducible $\mathcal{C}\ell^0_{p,q}$-module Σ is irreducible as module of the subalgebra $\mathcal{C}\ell^0_p \otimes \mathcal{C}\ell^0_{0,q} \subset \mathcal{C}\ell^0_{p,q} = \mathcal{C}\ell^0_p \otimes \mathcal{C}\ell^0_{0,q} + \mathcal{C}\ell^1_p \otimes \mathcal{C}\ell^1_{0,q}$ if $n \equiv 2, 4, 5$ or $6 \pmod 8$ and is the sum of two irreducible submodules if $n \equiv 0, 1, 3$ or $7 \pmod 8$. Now Lemma 2 follows from the fact that $\mathcal{C}\ell^0_p \otimes \mathcal{C}\ell^0_{0,q}$ (respectively, $\mathcal{C}\ell^0_{p,q}$) is the subalgebra of $\mathcal{C}\ell_{p,q}$ generated by 1 and $K = \mathrm{Spin}(p) \cdot \mathrm{Spin}(0, q)$ (respectively, by 1 and $\hat{K} = \mathrm{Spin}(p) \cdot \iota(\mathrm{Pin}(q))$). □

LEMMA 3. *A* $\mathcal{C}\ell^0_{p,q}$*-module* W *is irreducible if and only if it is irreducible as module of the maximal compact subgroup* $\hat{K} \subset \mathrm{Spin}(p, q)$. *In this case* $(\vee^2 W^*)^{\hat{K}}$ *is one-dimensional and is spanned by a positive definite scalar product on* W. *Let* W *be an irreducible* $\mathcal{C}\ell^0_{p,q}$*-module and* $\Pi : \wedge^2 W \to V$ *be any* $\mathfrak{o}(V)$*-equivariant linear map. Then either* $\Pi = 0$ *or* $b(\Pi)$ *is a definite* $\hat{K}$*-invariant symmetric bilinear form.*

Proof: The first statement follows from Lemma 2. Since $\hat{K}$ is compact there exists a positive definite $\hat{K}$-invariant symmetric bilinear form on W. From the irreducibility of W we conclude by Schur's lemma that $(\vee^2 W^*)^{\hat{K}}$ is spanned by this form. Now the last statement is an immediate consequence of Cor. 2. □

LEMMA 4. *Let* W *be a* $\mathcal{C}\ell^0(V)$*-module and* $\Pi : \wedge^2 W \to V$ *an* $\mathfrak{o}(V)$*-equivariant linear map such that* $b = b(\Pi)$ *is nondegenerate. Suppose that* $W = \Sigma \oplus \Sigma'$ *is the direct sum of two irreducible submodules* Σ *and* Σ'. *Then there exists a* b*-orthogonal decomposition* $W = \Sigma_1 \oplus \Sigma_2$ *into two* b*-nondegenerate (and hence* b*-definite by Lemma 3) irreducible submodules* Σ_1 *and* Σ_2.

Proof: It is sufficient to show that W contains a b-nondegenerate irreducible $\mathcal{C}\ell^0(V)$-submodule Σ_1. Then $\Sigma_2 := \Sigma_1^\perp$ is a b-nondegenerate $\hat{K}$-submodule. It is also a $\mathcal{C}\ell^0(V)$-submodule, because the algebra $\mathcal{C}\ell^0(V)$ is generated by 1 and $\hat{K}$, and it is irreducible since the $\mathcal{C}\ell^0(V)$-module W is the direct sum of only two irreducible submodules. If W does not contain any b-nondegenerate irreducible $\mathcal{C}\ell^0(V)$-submodule Σ_1 then, by Schur's lemma, the restriction of b to a bilinear form on any irreducible $\mathcal{C}\ell^0(V)$-submodule vanishes. In the following we derive a contradiction from this assumption. Since the bilinear form b is nondegenerate it defines a nondegenerate pairing between the b-isotropic subspaces Σ and Σ'. Due to the $\hat{K}$-invariance of b (Thm. 3) $b : \Sigma' \xrightarrow{\sim} \Sigma^*$ is a $\hat{K}$-equivariant isomorphism. On the other hand $\Sigma^* \cong \Sigma$ as irreducible modules of the compact group $\hat{K}$. This shows that Σ and Σ' are equivalent as $\hat{K}$-modules and thus as $C\ell^0(V)$-modules, because the algebra $C\ell^0(V)$ is generated by 1 and $\hat{K}$. Hence there exists a $C\ell^0(V)$-equivariant isomorphism $\varphi : \Sigma \xrightarrow{\sim} \Sigma'$. We define two $\hat{K}$-invariant bilinear forms $\beta_\pm$ on Σ by:

$$\beta_\pm(s,t) := b(\varphi(s),t) \pm b(\varphi(t),s)\,, \quad s,t \in \Sigma\,.$$

β_+ is symmetric and β_- is skew symmetric. If $\beta_+ \neq 0$ then it is a definite $\hat{K}$-invariant scalar product on Σ, since Σ is an irreducible module of the compact group $\hat{K}$. So for $s \in \Sigma - \{0\}$ we obtain

$$0 \neq \beta_+(s,s) = b(\varphi(s),s) + b(s,\varphi(s)) = b(s+\varphi(s), s+\varphi(s))\,.$$

This implies that the irreducible $C\ell^0(V)$-submodule $\Sigma_\varphi := \{s+\varphi(s) | s \in \Sigma\} \subset W$ is b-definite, which contradicts our assumption. We conclude that $\beta_+ = 0$ and hence $\beta_- = 2b \circ \varphi$ is a $\hat{K}$-invariant symplectic form on Σ. Let β be a $\hat{K}$-invariant positive definite scalar product on Σ (such scalar products exist since $\hat{K}$ is compact). We define a $\hat{K}$-equivariant isomorphism $\chi : \Sigma \to \Sigma$ by the equation

$$\beta(s,t) = \beta_-(\chi(s),t)\,, \quad s,t \in \Sigma\,.$$

Then $\psi := \varphi \circ \chi : \Sigma \to \Sigma'$ is a $\hat{K}$-equivariant and hence $C\ell^0(V)$-equivariant isomorphism. Using $\beta_+ = 0$, we compute for $s \in \Sigma - \{0\}$:

$$\begin{aligned} b(s+\psi(s), s+\psi(s)) &= b(\psi(s),s) + b(s,\psi(s)) = b(\psi(s),s) - b(\varphi(s),\chi(s)) \\ &= \beta_-(\chi(s),s) = \beta(s,s) \neq 0\,. \end{aligned}$$

As above, this implies that W contains a b-definite irreducible $C\ell^0(V)$-submodule $\Sigma_\psi \cong \Sigma$ contradicting our assumption. □

THEOREM 5. [**A-C2**] *Let* $V = \mathbb{R}^{p,q}$, $p \equiv 3 \pmod 4$. *If* W *is an irreducible* $C\ell^0(V)$*-module then*

$$\dim(\wedge^2 W^* \otimes V)^{\mathfrak{o}(V)} = 1\,.$$

Proof: Cor. 2 and Lemma. 3 show that $\dim(\wedge^2 W^* \otimes V)^{\mathfrak{o}(V)} \leq 1$. On the other hand, there exists a nontrivial $\mathfrak{o}(V)$-equivariant linear map $\wedge^2 W \to V$, see [**A-C2**], and hence $\dim(\wedge^2 W^* \otimes V)^{\mathfrak{o}(V)} \geq 1$. This proves that $\dim(\wedge^2 W^* \otimes V)^{\mathfrak{o}(V)} = 1$. □

Next, in order to parametrize the isomorphism classes of extended Poincaré algebras $\mathfrak{p}(\Pi) = \mathfrak{p}(V) + W$ we associate a certain number of nonnegative integers to $\mathfrak{p}(\Pi)$. Let us first consider the case when there is only one irreducible $C\ell^0(V)$-module Σ up to equivalence. Then W is necessarily isotypical and, due to Thm. 4, there exists a b-orthogonal decomposition $W = W_0 \oplus \oplus_{i=1}^l W_i$ with $W_0 = \ker b = \ker \Pi \cong l_0 \Sigma$ and irreducible $C\ell^0(V)$-submodules $W_i \cong \Sigma$ for $i = 1, \ldots, l$ on which b is definite. We denote by l_+ (repectively l_-) the number of summands W_i on which b is positive (respectively, negative) definite. Note that the triple (l_0, l_+, l_-) does not depend on the choice of decomposition.

THEOREM 6. *Let* $p \equiv 3 \pmod 4$ *and* $q \not\equiv 3 \pmod 4$. *Then the isomorphism class* $[\mathfrak{p}(\Pi)]$ *of an extended Poincaré algebra* $\mathfrak{p}(\Pi)$ *of signature* (p,q) *is completely determined by the triple* (l_0, l_+, l_-) *introduced above. We put* $\mathfrak{p}(p,q,l_0,l_+,l_-) := [\mathfrak{p}(\Pi)]$. *Then* $\mathfrak{p}(p,q,l_0,l_+,l_-) = \mathfrak{p}(p,q,l_0',l_+',l_-')$ *if and only if* $l_0 = l_0'$ *and* $\{l_+,l_-\} = \{l_+',l_-'\}$.

Proof: If $p \equiv 3 \pmod 4$ then there is only one irreducible $C\ell^0_{p,q}$-module Σ up to equivalence if and only if $q \not\equiv 3 \pmod 4$. Let $\mathfrak{p}(\Pi) = \mathfrak{p}(V) + W$ and $\mathfrak{p}(\Pi') = \mathfrak{p}(V) + W'$ be two extended Poincaré algebras of signature (p,q) with the same integers $l_0 = l_0(\Pi) = l_0(\Pi')$, $l_+ = l_+(\Pi) = l_+(\Pi')$ and $l_- = l_-(\Pi) = l_-(\Pi')$. Then the modules W and W' are equivalent and we can assume that $W = W' = W_0 \oplus \oplus_{i=1}^{l=l_++l_-} W_i$ is a decomposition as above. In particular, it is $b(\Pi)$- and $b(\Pi')$-orthogonal, $b(\Pi)$ and $b(\Pi')$ are both positive definite or both negative definite on W_i for $i \geq 1$, $\Pi(W_0 \wedge W) = \Pi'(W_0 \wedge W) = 0$, $\Pi(W_i \wedge W_j) = \Pi'(W_i \wedge W_j) = 0$ if $i \neq j$ and $\Pi(\wedge^2 W_i) = \Pi'(\wedge^2 W_i) = V$ if $i \geq 1$. So the maps Π and Π' are completely determined by their restrictions $\Pi_i := \Pi| \wedge^2 W_i \neq 0$ and $\Pi'_i := \Pi'| \wedge^2 W_i \neq 0$ $(i \geq 1)$ respectively. By Thm. 5 $\Pi'_i = \lambda_i \Pi_i$ $(i \geq 1)$ for some constant $\lambda_i \in \mathbb{R}^*$. Now $b = b(\Pi)$ and $b' = b(\Pi')$ are both positive definite or both negative definite on W_i and hence $\lambda_i = \mu_i^2 > 0$. Now we can define an isomorphism $\varphi : \mathfrak{p}(\Pi) \to \mathfrak{p}(\Pi')$ by $\varphi|\mathfrak{p}(V) + W_0 = \mathrm{Id}$ and $\varphi|W_i = \mu_i \mathrm{Id}$. This shows that the integers (l_0, l_+, l_-) determine the extended Poincaré algebra $\mathfrak{p}(\Pi)$ of signature (p,q) up to isomorphism. The $\mathbb{Z}_2$-graded Lie algebras $\mathfrak{p}(\Pi)$ and $\mathfrak{p}(-\Pi)$ are isomorphic via $\alpha : \mathfrak{p}(\Pi) \to \mathfrak{p}(-\Pi)$ defined by: $\alpha|\mathfrak{o}(V) + W = \mathrm{Id}$ and $\alpha|V = -\mathrm{Id}$. This proves $\mathfrak{p}(p,q,l_0,l_+,l_-) = \mathfrak{p}(p,q,l_0,l_-,l_+)$. It remains to show that $\mathfrak{p}(p,q,l_0,l_+,l_-) = \mathfrak{p}(p,q,l'_0,l'_+,l'_-)$ implies $l_0 = l'_0$ and $\{l_+, l_-\} = \{l'_+, l'_-\}$. Let $\mathfrak{p}(\Pi) = \mathfrak{p}(V) + W \in \mathfrak{p}(p,q,l_0,l_+,l_-)$ and $\mathfrak{p}(\Pi') = \mathfrak{p}(V) + W' \in \mathfrak{p}(p,q,l'_0,l'_+,l'_-)$ be representative extended Poincaré algebras and $W = W_0 \oplus \oplus_{i=1}^{l} W_i$ $(l = l_+ + l_-)$ a decomposition as above. We assume that there exists an isomorphism $\varphi : \mathfrak{p}(\Pi) \to \mathfrak{p}(\Pi')$ of $\mathbb{Z}_2$-graded Lie algebras, i.e. $\varphi\mathfrak{p}(V) = \mathfrak{p}(V)$ and $\varphi W = W$. The automorphism $\varphi|\mathfrak{p}(V)$ preserves the radical V and maps the Levi subalgebra $\mathfrak{o}(V)$ to an other Levi subalgebra of $\mathfrak{p}(V)$. Now by Malcev's theorem any two Levi subalgebras are conjugated by an inner automorphism, see [**O-V**]. So, using an inner automorphism of $\mathfrak{p}(\Pi)$, we can assume that $\varphi\mathfrak{o}(V) = \mathfrak{o}(V)$. The subalgebra $\varphi(\mathfrak{o}(p) \oplus \mathfrak{o}(q)) \subset \mathfrak{o}(V)$ is maximal compact (i.e. the Lie algebra of a maximal compact subgroup of $\mathrm{O}(V) = \mathrm{O}(p,q)$) and hence conjugated by an inner automorphism to the maximal compact subalgebra $\mathfrak{o}(p) \oplus \mathfrak{o}(q) \subset \mathfrak{o}(V)$. So, again, we can assume that φ preserves $\mathfrak{o}(p) \oplus \mathfrak{o}(q)$. Moreover, since $p \neq q$ any automorphism of $\mathfrak{o}(p) \oplus \mathfrak{o}(q)$ is inner and we can assume that $\varphi|\mathfrak{o}(p) \oplus \mathfrak{o}(q) = \mathrm{Id}$. From the fact that φ is an automorphism of $\mathfrak{p}(V)$ we obtain that $\phi := \varphi|V \in \mathrm{GL}(V)$ normalizes $\mathfrak{o}(V)$ and $\mathfrak{o}(p) \oplus \mathfrak{o}(q)$. This implies $\phi \in \mathrm{O}(p) \times \mathrm{O}(q)$, in particular, $\varphi\mathbb{R}^p = \mathbb{R}^p$ and $\varphi\mathbb{R}^{0,q} = \mathbb{R}^{0,q}$. Using an inner automorphism of $\mathfrak{p}(\Pi)$ we can further assume that $\varphi|\mathbb{R}^p = \epsilon(\varphi)\mathrm{Id}$, $\epsilon(\varphi) \in \{+1, -1\}$, and hence $\varphi|\mathfrak{o}(p) = \mathrm{Ad}_\phi|\mathfrak{o}(p) = \mathrm{Id}$. This implies that $\varphi|W$ is $\mathrm{Spin}(p)$-equivariant. Now we can compute

$$\begin{aligned} b'(\varphi s, \varphi t) &= \langle e_1, [e_2 \dots e_p \varphi s, \varphi t] \rangle = \langle e_1, [\varphi e_2 \dots e_p s, \varphi t] \rangle \\ &= \langle e_1, \varphi [e_2 \dots e_p s, t] \rangle = \epsilon(\varphi) \langle e_1, [e_2 \dots e_p s, t] \rangle \\ &= \epsilon(\varphi) b(s,t)\,, \quad s, t \in W\,. \end{aligned}$$

Put $W'_i := \varphi W_i$. Then, since $\varphi^* b' = \epsilon(\varphi) b$, we obtain a b'-orthogonal decomposition $W' = W'_0 \oplus \oplus_{i=1}^{l'} W'_i$ $(l' = l'_+ + l'_-)$ as above; $W'_0 = \ker \Pi' \cong l'_0 \Sigma$, $W'_i \cong \Sigma$ $(i \geq 1)$ etc. This shows that $l'_0 = l_0$, $l'_\pm = l_\pm$ if $\epsilon(\varphi) = +1$ and $l'_\pm = l_\mp$ if $\epsilon(\varphi) = -1$. □

Now we discuss the complementary case $p \equiv q \equiv 3 \pmod 4$. In this case, the spinor module S of $C\ell^0(V)$ is the sum $S^+ \oplus S^-$ of two irreducible inequivalent semispinor modules S^+ and S^- and any irreducible $C\ell^0(V)$-module is equivalent to S^+ or S^-. As $\mathrm{Spin}_0(V)$-modules, S^+ and S^- are dual: $S^- \cong (S^+)^*$. Let

$\mathfrak{p}(\Pi) = \mathfrak{p}(V) + W$ be an extended Poincaré algebra of signature (p,q) as above. Thanks to Thm. 4 there exists a b-orthogonal decomposition $W = W_0 \oplus \oplus_{i=1}^{l} W_i$ as above with the following b-orthogonal refinements: $W_0 = W_0^+ \oplus W_0^-$, $W_0^\pm \cong l_0^\pm S^\pm$, and $\oplus_{i=1}^{l} W_i = \oplus_{i=1}^{l^+} W_i^+ \oplus \oplus_{i=1}^{l^-} W_i^-$, $W_i^\pm \cong S^\pm$. We denote by l_+^ϵ (respectively, l_-^ϵ) the number of submodules W_i^ϵ, $i = 1, \ldots, l^\epsilon$, on which b is positive (respectively, negative) definite, $\epsilon \in \{+,-\}$. So to $\mathfrak{p}(\Pi)$ we have associated the nonnegative integers $(l_0^+, l_+^+, l_-^+, l_0^-, l_+^-, l_-^-)$.

THEOREM 7. *Let* $p \equiv q \equiv 3 \pmod 4$. *Then the isomorphism class* $[\mathfrak{p}(\Pi)]$ *of an extended Poincaré algebra* $\mathfrak{p}(\Pi)$ *of signature* (p,q) *is completely determined by the tuple* $(l_0^+, l_+^+, l_-^+, l_0^-, l_+^-, l_-^-)$ *introduced above. We put* $\mathfrak{p}(p,q,l_0^+, l_+^+, l_-^+, l_0^-, l_+^-, l_-^-) := [\mathfrak{p}(\Pi)]$. *Then* $\mathfrak{p}(p,q,l_0^+, l_+^+, l_-^+, l_0^-, l_+^-, l_-^-) = \mathfrak{p}(p,q,\tilde{l}_0^+, \tilde{l}_+^+, \tilde{l}_-^+, \tilde{l}_0^-, \tilde{l}_+^-, \tilde{l}_-^-)$ *if and only if* $(\tilde{l}_0^+, \tilde{l}_+^+, \tilde{l}_-^+, \tilde{l}_0^-, \tilde{l}_+^-, \tilde{l}_-^-) \in \Gamma(l_0^+, l_+^+, l_-^+, l_0^-, l_+^-, l_-^-)$, *where* $\Gamma \cong \mathbb{Z}_2 \times \mathbb{Z}_2$ *is the group generated by the following two involutions:*

$$(l_0^+, l_+^+, l_-^+, l_0^-, l_+^-, l_-^-) \mapsto (l_0^+, l_-^+, l_+^+, l_0^-, l_-^-, l_+^-)$$

and

$$(l_0^+, l_+^+, l_-^+, l_0^-, l_+^-, l_-^-) \mapsto (l_0^-, l_+^-, l_-^-, l_0^+, l_+^+, l_-^+)\,.$$

Proof: The proof uses again Thm. 4 and Thm. 5 and is similar to that of Thm. 6. Therefore, we explain only the reason for the appearance of the second involution. In terms of the standard basis $(e_1, \ldots, e_p, e_1', \ldots, e_q')$ of $V = \mathbb{R}^{p,q}$ we define an isometry $\phi \in \mathrm{SO}(p) \times \mathrm{O}(q)$ by: $\phi e_1 := e_1$, $\phi e_i := -e_i$ $(i \geq 2)$ and $\phi e_j' = -e_j'$ $(j \geq 1)$. Then $\mathrm{Ad}_\phi \in \mathrm{Aut}\,\mathfrak{p}(V)$ induces an (outer) automorphism of $\mathfrak{o}(V)$ interchanging the two semispinor modules. Let $(\mathfrak{p}(V) + W, [\cdot,\cdot]) \in \mathfrak{p}(p,q,l_0^+, l_+^+, l_-^+, l_0^-, l_+^-, l_-^-)$ be an extended Poincaré algebra of signature (p,q). Then we define a new extended Poincaré algebra $(\mathfrak{p}(V) + W, [\cdot,\cdot]') \in \mathfrak{p}(p,q,l_0^-, l_+^-, l_-^-, l_0^+, l_+^+, l_-^+)$ by:

$$[\cdot,\cdot]' := [\cdot,\cdot] \quad \text{on} \quad \wedge^2 \mathfrak{p}(V) \oplus \wedge^2 W \oplus V \wedge W$$

and

$$[A,s]' := [\mathrm{Ad}_\phi(A), s] \quad \text{for} \quad A \in \mathfrak{o}(V)\,, s \in W\,.$$

The two $\mathbb{Z}_2$-graded Lie algebras are isomorphic via $\varphi : (\mathfrak{p}(V) + W, [\cdot,\cdot]) \xrightarrow{\sim} (\mathfrak{p}(V) + W, [\cdot,\cdot]')$ defined by: $\varphi|\mathfrak{p}(V) = \mathrm{Ad}_\phi$ and $\varphi|W = \mathrm{Id}$. $\square$

2. The homogeneous quaternionic manifold (M, Q) associated to an extended Poincaré algebra

2.1. Homogeneous manifolds associated to extended Poincaré algebras. Any extended Poincaré algebra $\mathfrak{p} = \mathfrak{p}(\Pi) = \mathfrak{p}(V) + W$ has an even derivation D with eigenspace decomposition $\mathfrak{p} = \mathfrak{o}(V) + V + W$ and corresponding eigenvalues $(0, 1, 1/2)$. Therefore, the $\mathbb{Z}_2$-graded Lie algebra $\mathfrak{p} = \mathfrak{p}_0 + \mathfrak{p}_1 = \mathfrak{p}(V) + W$ is canonically extended to a $\mathbb{Z}_2$-graded Lie algebra $\mathfrak{g} = \mathfrak{g}(\Pi) = \mathbb{R}D + \mathfrak{p} = \mathfrak{g}_0 + \mathfrak{g}_1$, where $\mathfrak{g}_0 = \mathbb{R}D + \mathfrak{p}_0 = \mathbb{R}D + \mathfrak{p}(V) =: \mathfrak{g}(V)$ and $\mathfrak{g}_1 = \mathfrak{p}_1 = W$. The next proposition describes the basic structure of the Lie algebra $\mathfrak{g} = \mathfrak{g}(\Pi) = \mathfrak{g}(V) + W$. To avoid trivial exceptions, in the following we assume that $n = \dim V > 2$ and hence that $\mathfrak{o}(V)$ is semisimple (for the construction of homogeneous quaternionic manifolds we will put $V = \mathbb{R}^{p,q}$ and $p \geq 3$).

PROPOSITION 2. *The Lie algebra $\mathfrak{g} = \mathfrak{g}(\Pi)$ has the Levi decomposition*

$$\mathfrak{g} = \mathfrak{o}(V) + \mathfrak{r} \tag{6}$$

into the radical $\mathfrak{r} = \mathbb{R}D + V + W$ and the complementary (maximal semisimple) Levi subalgebra $\mathfrak{o}(V)$. The nilradical $\mathfrak{n} = [\mathfrak{r}, \mathfrak{r}] = V + W$ is two-step nilpotent if $\Pi \neq 0$ and Abelian otherwise.

For any Lie algebra $\mathfrak{l}$ we denote by $\mathrm{der}(\mathfrak{l})$ the Lie algebra of its derivations.

PROPOSITION 3. *The adjoint representation $\mathfrak{g} \to \mathrm{der}(\mathfrak{r})$ of $\mathfrak{g} = \mathfrak{o}(V) + \mathfrak{r}$ on its ideal $\mathfrak{r} = \mathbb{R}D + V + W$ is faithful.*

Proof: Let $x \in \mathfrak{g}$. We show that $[x, \mathfrak{r}] = 0$ implies $x = 0$. First $[x, D] = 0$ implies $x \in \mathfrak{co}(V) = \mathbb{R}D + \mathfrak{o}(V)$. Then $[x, V] = 0$ implies $x = 0$, because the conformal Lie algebra $\mathfrak{co}(V)$ acts faithfully on V. $\square$

By Prop. 3 we can consider $\mathfrak{g}(V)$ and $\mathfrak{g} = \mathfrak{g}(\Pi) = \mathfrak{g}(V) + W$ as linear Lie algebras via the embedding $\mathfrak{g}(V) \subset \mathfrak{g} \hookrightarrow \mathrm{der}(\mathfrak{r}) \subset \mathfrak{gl}(\mathfrak{r})$. The corresponding connected Lie groups of $\mathrm{Aut}(\mathfrak{r}) \subset \mathrm{GL}(\mathfrak{r})$ will be denoted by $G(V)$ and $G = G(\Pi)$ respectively: $\mathrm{Lie}\, G(V) = \mathfrak{g}(V) \subset \mathrm{Lie}\, G = \mathfrak{g}$. Now let $V = \mathbb{R}^{p,q}$ and fix a decomposition $p = p' + p''$. The subalgebra of $\mathfrak{o}(V)$ preserving the corresponding orthogonal splitting $V = E + E' = \mathbb{R}^{p',0} + \mathbb{R}^{p'',q}$ is $\mathfrak{k} = \mathfrak{k}(p', p'') = \mathfrak{o}(p') \oplus \mathfrak{o}(p'', q) \subset \mathfrak{o}(p, q) = \mathfrak{o}(V)$. We consider $\mathfrak{k} \subset \mathfrak{o}(V) \subset \mathfrak{g}$ as a subalgebra of the linear Lie algebra $\mathfrak{g} \hookrightarrow \mathrm{der}(\mathfrak{r})$ and denote by $K = K(p', p'') \subset G(V) \subset G \subset \mathrm{Aut}(\mathfrak{r})$ the corresponding connected linear Lie group. K is a closed Lie subgroup, see Cor. 3 below. We are interested in the homogeneous spaces

$$M(V) := G(V)/K \subset M = M(\Pi) := G/K = G(\Pi)/K\,.$$

PROPOSITION 4. *The Lie subalgebras $\mathfrak{k}$, $\mathfrak{n}$, $\mathfrak{r}$, $\mathfrak{o}(V)$, $\mathfrak{p}(V)$, $\mathfrak{p}$, $\mathfrak{g}(V)$, $\mathfrak{g} \subset \mathrm{der}(\mathfrak{r})$ are algebraic subalgebras of the (real) algebraic Lie algebra $\mathrm{der}(\mathfrak{r})$.*

Proof: We use the following sufficient conditions for algebraicity, see [**O-V**] Ch. 3 §3 8°:

a) A linear Lie algebra coinciding with its derived algebra is algebraic.
b) The radical of an algebraic linear Lie algebra is algebraic.
c) A linear Lie algebra generated by algebraic subalgebras is algebraic.

The subalgebras $\mathfrak{o}(V) \subset \mathfrak{p}(V) \subset \mathfrak{p} \subset \mathrm{der}(\mathfrak{r})$ are algebraic by a). By c), to prove that $\mathfrak{g} = \mathbb{R}D + \mathfrak{p}$ and $\mathfrak{g}(V)$ are algebraic it is now sufficient to show that $\mathbb{R}D \subset \mathrm{der}(\mathfrak{r})$ is algebraic. D preserves $\mathfrak{n} = V + W$ and acts trivially on the complement $\mathbb{R}D + \mathfrak{o}(V)$. The Lie algebra $\mathbb{R}D \hookrightarrow \mathrm{der}(\mathfrak{n})$ is the Lie algebra of the algebraic group

$$\{\lambda \mathrm{Id}_V \oplus \mu \mathrm{Id}_W | \lambda = \mu^2 \neq 0\} \subset \mathrm{GL}(V \oplus W).$$

This shows that $\mathbb{R}D$, $\mathfrak{g}(V)$, $\mathfrak{g} \subset \mathrm{der}(\mathfrak{r})$ are algebraic. Now $\mathfrak{n}$, the radical of $\mathfrak{p}$, and $\mathfrak{r}$, the radical of $\mathfrak{g}$, are algebraic by b). Finally, $\mathfrak{k} \subset \mathfrak{o}(V)$ is the subalgebra which preserves the orthogonal splitting $V = E + E'$ and is hence algebraic. □

COROLLARY 3. *The connected linear Lie groups K, $S \cong \mathrm{Spin}_0(V)$, $R = \exp \mathfrak{r}$, $G(V)$, $G \subset \mathrm{Aut}(\mathfrak{r})$ with Lie algebras $\mathfrak{k}$, $\mathfrak{s} := \mathfrak{o}(V)$, $\mathfrak{r}$, $\mathfrak{g}(V)$, $\mathfrak{g} \subset \mathrm{der}(\mathfrak{r})$ are closed.*

Proof: This follows from Prop. 4 and the fact that the identity component of a real algebraic linear group is a closed linear group. □

PROPOSITION 5. *The Lie group $G(\Pi)$ has the following Levi decomposition:*

$$G(\Pi) = S \ltimes R. \tag{7}$$

Proof: This follows from the corresponding Levi decomposition (6) of Lie algebras since $S \cap R = \{e\}$. □

Next we show that $M(V)$ can be naturally endowed with a $G(V)$-invariant structure of pseudo-Riemannian locally symmetric space. With this in mind, we consider the pseudo-Euclidean vector space $\tilde{V} = \mathbb{R}^{p+1,q+1}$ with scalar product $\langle \cdot, \cdot \rangle$ and orthogonal decomposition $\tilde{V} = \mathbb{R}e_0 + V + \mathbb{R}e_0'$, $\langle e_0, e_0 \rangle = -\langle e_0', e_0' \rangle = 1$. Recall that $\mathfrak{o}(\tilde{V})$ is identified with $\wedge^2 \tilde{V}$ via the pseudo-Euclidean scalar product $\langle \cdot, \cdot \rangle$, see (3).

PROPOSITION 6. *The subspace*

$$\mathbb{R}e_0 \wedge e_0' + \wedge^2 V + (e_0 - e_0') \wedge V \subset \wedge^2 \tilde{V} = \mathfrak{o}(\tilde{V})$$

is a subalgebra isomorphic to $\mathfrak{g}(V)$.

Proof: The canonical embedding $\mathfrak{o}(V) = \wedge^2 V \hookrightarrow \wedge^2 \tilde{V} = \mathfrak{o}(\tilde{V})$ is extended to an embedding $\mathfrak{g}(V) \hookrightarrow \mathfrak{o}(\tilde{V})$ via

$$D \mapsto e_0 \wedge e_0', \quad V \ni v \mapsto (e_0 - e_0') \wedge v. \square$$

It is easy to see that the embedding $\mathfrak{g}(V) \hookrightarrow \mathfrak{o}(\tilde{V})$ lifts to a homomorphism of connected Lie groups $G(V) \to \mathrm{SO}_0(\tilde{V})$ with finite kernel. In particular, we have a natural isometric action of $G(V)$ on the pseudo-Riemannian symmetric space $\tilde{M}(V) := \mathrm{SO}_0(p+1, q+1)/(\mathrm{SO}(p'+1) \times \mathrm{SO}_0(p'', q+1))$. We denote by $[e] = e\tilde{K} \in \tilde{M}(V) := \mathrm{SO}_0(\tilde{V})/\tilde{K}$, $\tilde{K} := \mathrm{SO}(p'+1) \times \mathrm{SO}_0(p'', q+1)$, the canonical base point and by $G(V)[e]$ its $G(V)$-orbit.

PROPOSITION 7. *The action of $G(V)$ on $\tilde{M}(V)$ induces a $G(V)$-equivariant open embedding*

$$M(V) = G(V)/K \xrightarrow{\sim} G(V)[e] \subset \tilde{M}(V).$$

$G(V)$ acts transitively on $\tilde{M}(V)$ if and only if $\tilde{M}(V)$ is Riemannian, i.e. if and only if $p'' = 0$. In that case $M(V) \cong \tilde{M}(V)$ is the noncompact dual of the Grassmannian $\mathrm{SO}(p+q+2)/(\mathrm{SO}(p+1) \times \mathrm{SO}(q+1))$ and admits a simply transitive splittable solvable subgroup $I(S) \ltimes \exp(\mathbb{R}D + V) \subset G(V)$. Here $I(S)$ denotes the (solvable) Iwasawa subgroup of $S \cong \mathrm{Spin}_0(V)$.

Proof: The stabilizer $G(V)_{[e]}$ of $[e]$ in $G(V)$ coincides with K and hence $G(V)[e] \cong G(V)/G(V)_{[e]} = G(V)/K = M(V)$. Now a simple dimension count shows that the orbit $G(V)[e] \subset \tilde{M}(V)$ is open. If $\tilde{M}(V)$ is Riemannian then it is a Riemannian symmetric space of noncompact type and, by the Iwasawa decomposition theorem (see [**H**]) the Iwasawa subgroup $I(\mathrm{SO}_0(p+1, q+1)) \subset \mathrm{SO}_0(p+1, q+1)$ is a splittable solvable subgroup which acts simply transitively on $\tilde{M}(V)$. Now let $G(V) = S \ltimes R(V)$ be the Levi decomposition associated to the Levi decomposition $\mathfrak{g}(V) = \mathfrak{o}(V) + (\mathbb{R}D + V)$, $S \cong \mathrm{Spin}_0(V)$, $R(V) = \exp(\mathbb{R}D + V)$ (cf. Prop. 2). We denote by $I(S) \subset S$ the Iwasawa subgroup of S. Then $I(S) \ltimes R(V) \subset G(V)$ is mapped isomorphically onto $I(\mathrm{SO}_0(p+1, q+1)) \subset \mathrm{SO}_0(\tilde{V})$ by the homomorphism $G(V) \to \mathrm{SO}_0(\tilde{V})$ introduced above. This shows that $I(S) \ltimes R(V)$ and hence $G(V)$ acts transitively on $\tilde{M}(V)$ in the Riemannian case.

If $\tilde{M}(V)$ is not Riemannian then the homogeneous spaces $\tilde{M}(V)$ and $M(V)$ are not homotopy equivalent and hence $G(V)$ does not act transitively on $\tilde{M}(V)$. In fact, $\tilde{M}(V)$ (respectively, $M(V)$) has the homotopy type of the Grassmanian $\mathrm{SO}(p+1)/(\mathrm{SO}(p'+1) \times \mathrm{SO}(p''))$ (respectively, $\mathrm{SO}(p)/(\mathrm{SO}(p') \times \mathrm{SO}(p''))$). Now it is sufficient to observe that the Stiefel manifolds $\mathrm{SO}(p+1)/\mathrm{SO}(p'+1)$ and $\mathrm{SO}(p)/\mathrm{SO}(p')$ are homotopy equivalent only if $p = p'$ and hence $p'' = 0$, see [**O3**]. $\square$

Before we can formulate the main result of the present paper in section 2.4 we need to recall the notions of quaternionic manifold and of (pseudo-) quaternionic Kähler manifold, cf. [**A-M2**]. The reader familiar with these concepts should skip the next section.

2.2. Quaternionic structures. It is instructive to introduce the basic concepts of quaternionic geometry as analogues of the more familiar concepts of complex geometry.

DEFINITION 4. *Let E be a (finite dimensional) real vector space. A* **complex structure** *on E is an endomorphism $J \in \mathrm{End}(E)$ such that $J^2 = -\mathrm{Id}$. A* **hypercomplex structure** *on E is a triple $(J_\alpha) = (J_1, J_2, J_3)$ of complex structures on E satisfying $J_1 J_2 = J_3$. A* **quaternionic structure** *on E is the three-dimensional subspace $Q \subset \mathrm{End}(E)$ spanned by a hypercomplex structure (J_α): $Q = \mathrm{span}\{J_1, J_2, J_3\}$. In that case, we say that the hypercomplex structure (J_α) is* **subordinate** *to the quaternionic structure Q.*

Note, first, that $Q \subset \mathfrak{gl}(E)$ is a Lie subalgebra isomorphic to $\mathfrak{sp}(1) \cong \mathrm{Im}\mathbb{H} = \mathrm{span}\{i, j, k\}$ the Lie algebra of the group $Sp(1) = S^3 \subset \mathbb{H} = \mathrm{span}\{1, i, j, k\}$ of unit quaternions and, second, that the real associative subalgebra of $\mathrm{End}(E)$ generated by Id and a quaternionic structure Q on E is isomorphic to the algebra of quaternions $\mathbb{H}$. In both cases, the choice of such isomorphism is equivalent to the choice of a hypercomplex structure $(J_\alpha) = (J_1, J_2, J_3)$ subordinate to Q. In fact, given (J_α) we can define an isomorphism of associative algebras by $(\mathrm{Id}, J_1, J_2, J_3) \mapsto (1, i, j, k)$ this induces also an isomorphism of Lie algebras $Q \xrightarrow{\sim} \mathfrak{sp}(1)$.

DEFINITION 5. *Let M be a (smooth) manifold. An* **almost complex structure** *J (respectively,* **almost hypercomplex structure** *(J_1, J_2, J_3),* **almost quaternionic structure** *Q) on M is a (smooth) field $M \ni m \mapsto J_m \in \mathrm{End}(T_m M)$ of complex structures (respectively, $m \mapsto (J_1, J_2, J_3)_m$ of hypercomplex structures, $m \mapsto Q_m$ of quaternionic structures). The pair (M, J) (respectively, $(M, (J_\alpha))$, (M, Q)) is called an* **almost complex manifold** *(respectively,* **almost hypercomplex manifold, almost quaternionic manifold***).*

We recall that a connection on a manifold M (i.e. a covariant derivative ∇ on its tangent bundle TM) induces a covariant derivative ∇ on the full tensor algebra over TM and, in particular, on $\mathrm{End}(TM) \cong TM \otimes T^*M$. We will say that ∇ preserves a subbundle $B \subset \mathrm{End}(TM)$ if it maps (smooth) sections of B into sections of $T^*M \otimes B$.

DEFINITION 6. *A connection ∇ on an almost complex manifold (M, J) (respectively, almost hypercomplex manifold $(M, (J_\alpha))$, almost quaternionic manifold (M, Q)) is called* **almost complex** *(respectively,* **almost hypercomplex, almost quaternionic***)* **connection** *if $\nabla J = 0$ (respectively, if $\nabla J_1 = \nabla J_2 = \nabla J_3 = 0$, if ∇ preserves the rank 3 subbundle $Q \subset \mathrm{End}(TM)$). A* **complex** *(respectively,* **hypercomplex, quaternionic***)* **connection** *is a torsionfree almost complex (respectively, almost hypercomplex, almost quaternionic) connection.*

Note that the equation $\nabla J = 0$ is equivalent to the condition that ∇ preserves the rank 1 subbundle of $\mathrm{End}(TM)$ spanned by J, i.e. $\nabla J = \theta \otimes J$ for some 1-form θ on M. Therefore, the notion of almost quaternionic connection (as well as that of almost hypercomplex connection) is a direct quaternionic analogue of the notion of almost complex connection.

DEFINITION 7. *Let M be a manifold. An almost complex structure (respectively, almost hypercomplex structure, almost quaternionic structure) on M is called* **1-integrable** *if there exists a complex (respectively, hypercomplex, quaternionic) connection on M. A* **complex structure** *(respectively,* **hypercomplex structure, quaternionic structure***) on M is a 1-integrable almost complex structure (respectively, almost hypercomplex structure, almost quaternionic structure) on M.*

Remark 3: It is well known, see [**N-N**] and [**K-NII**], that an almost complex manifold (M, J) is integrable, i.e. admits an atlas with holomorphic transition maps, if and only if it is 1-integrable. This justifies the following definition of complex manifold.

DEFINITION 8. *A* **complex manifold** *(respectively,* **hypercomplex manifold***) is a manifold M together with a complex structure J (respectively, hypercomplex structure (J_α)) on M. A* **quaternionic manifold** *of dimension $d > 4$ is a manifold M of dimension d together with a quaternionic structure Q on M. Finally, a* **quaternionic manifold** *of dimension $d = 4$ is a 4-dimensional manifold M together with an almost quaternionic structure Q which annihilates the Weyl tensor of the conformal structure defined by Q, see Remark 4 below.*

For examples of hypercomplex and quaternionic manifolds (without metric condition) see [**J1**], [**J2**] and [**B-D**].

Remark 4: Notice that an almost quaternionic structure on a 4-manifold induces an oriented conformal structure. This follows from the fact that the normalizer in $\mathrm{GL}(4, \mathbb{R})$ of the standard quaternionic structure(s) on $\mathbb{R}^4 = \mathbb{H}$ is the special conformal group $\mathrm{CO}_0(4)$. The definition of quaternionic 4-manifold (M, Q) implies that this conformal structure is half-flat. More precisely, if the orientation of M is chosen such that Q is locally generated by positively oriented almost complex structures, then the self-dual half of the Weyl tensor vanishes. The special treatment of the 4-dimensional case in Def. 8 and also in Def. 11 below has the advantage that with these definitions many important properties of quaternionic (and also of quaternionic Kähler) manifolds of dimension > 4 remain true in dimension 4, cf.

Remark 5. In particular, all future statements about quaternionic manifolds and quaternionic Kähler manifolds in the present paper, such as the integrability of the canonical almost complex structure on the twistor space (see section 3), are valid also in the 4-dimensional case.

Next we discuss (almost) complex, hypercomplex and quaternionic structures on a pseudo-Riemannian manifold (M, g).

DEFINITION 9. *A pseudo-Riemannian metric g on an almost complex manifold (M, J) (respectively, almost hypercomplex manifold $(M, (J_\alpha))$, almost quaternionic manifold (M, Q)) is called* **Hermitian** *if J is skew symmetric (respectively, the J_α are skew symmetric, Q consists of skew symmetric endomorphisms).*

Note that, due to $J^2 = -\mathrm{Id}$, an almost complex structure J on a pseudo-Riemannian manifold (M, g) is skew symmetric if and only if J is orthogonal, i.e. if and only if $g(JX, JY) = g(X, Y)$ for all vector fields X, Y on M. Similarly, an almost quaternionic structure Q on a pseudo-Riemannian manifold (M, g) consists of skew symmetric endomorphisms if and only if $Z := \{A \in Q | A^2 = -\mathrm{Id}\}$ consists of orthogonal endomorphisms (here the equation $A^2 = -\mathrm{Id}$ is on $T_{\pi A}M$, $\pi : Q \to M$ the bundle projection).

DEFINITION 10. *A* **complex** *(pseudo-)* **Hermitian manifold** *(M, J, g) (respectively,* **hypercomplex** *(pseudo-)* **Hermitian manifold** *$(M, (J_\alpha), g)$,* **quaternionic** *(pseudo-)* **Hermitian manifold** *(M, Q, g),* **almost complex** *(pseudo-)* **Hermitian manifold** *(M, J, g) etc.) is a complex manifold (M, J) (respectively, hypercomplex manifold $(M, (J_\alpha))$, quaternionic manifold (M, Q), almost complex manifold (M, J) etc.) with a Hermitian (pseudo-) Riemannian metric g.*

Next we define the hypercomplex and quaternionic analogues of (pseudo-) Kähler manifolds.

DEFINITION 11. *A (pseudo-)* **Kähler manifold** *(respectively, (pseudo-)* **hyper-Kähler manifold**, **quaternionic** *(pseudo-)* **Kähler manifold** *of dimension $d > 4$) is an almost complex (pseudo-) Hermitian manifold (M, J, g) (respectively, almost hypercomplex (pseudo-) Hermitian manifold $(M, (J_\alpha), g)$, almost quaternionic (pseudo-) Hermitian manifold (M, Q, g) of dimension $d > 4$) with the property that the Levi-Civita connection ∇^g of the (pseudo-) Riemannian metric g is complex (respectively, hypercomplex, quaternionic). An almost quaternionic Hermitian 4-manifold (M, Q, g) is called* **quaternionic Kähler manifold** *if Q annihilates the curvature tensor R of ∇^g.*

See [**Bes**], [**H-K-L-R**], [**C-F-G**], [**Sw1**], [**D-S1**], [**D-S2**], [**Bi1**], [**Bi2**], [**K-S2**] and [**C4**] for examples of hyper-Kähler manifolds and [**W1**], [**A3**], [**Bes**], [**G1**], [**G-L**], [**L2**], [**L3**], [**F-S**], [**dW-VP2**], [**A-G**], [**K-S1**], [**A-P**], [**D-S3**] and [**C2**] for examples of quaternionic Kähler manifolds.

Remark 5: As explained in Remark 4, an almost quaternionic structure Q on a 4-manifold M defines a conformal structure. It is clear that a pseudo-Riemannian metric g on M defines the same conformal structure as Q if and only if it is Q-Hermitian and that any such metric is definite. Moreover, the Levi-Civita connection ∇^g of an almost quaternionic Hermitian 4-manifold (M, Q, g) automatically preserves Q. In fact, its holonomy group at $m \in M$ is a subgroup of $\mathrm{SO}(T_m M, g_m)$ because M is oriented (by Q) and the latter group normalizes Q because g is Q-Hermitian. Now let (M, Q, g) be a quaternionic Kähler 4-manifold. Since Q

annihilates the curvature tensor R and the metric g it must also annihilate the the Ricci tensor Ric and the Weyl tensor of (M, g). This shows first that Ric is Q-Hermitian and hence proportional to g: $Ric = cg$. In other words, (M, g) is an Einstein manifold. Second, (M, Q) is a quaternionic manifold because Q annihilates the Weyl tensor of the conformal structure defined by g, which coincides with the conformal structure defined by Q. For any quaternionic pseudo-Kähler manifold (M, g) (of arbitrary dimension) it is known, see e.g. [**A-M2**], that (M, g) is Einstein and that Q annihilates R.

As next, we introduce the appropriate notions in order to discuss transitive group actions on manifolds with the special geometric structures defined above.

2.3. Invariant connections on homogeneous manifolds and 1-integrability of homogeneous almost quaternionic manifolds.

DEFINITION 12. *The* **automorphism group** *of an almost complex manifold (M, J) (respectively, almost hypercomplex manifold $(M, (J_\alpha))$, almost quaternionic manifold (M, Q), almost complex Hermitian manifold (M, J, g), etc.) is the group of diffeomorphisms of M which preserves J (respectively, (J_α), Q, (J, g), etc.). An almost complex manifold (respectively, almost hypercomplex manifold, almost quaternionic manifold, almost complex Hermitian manifold, etc.) is called* **homogeneous** *if it has a transitive automorphism group.*

In the next section, we are going to construct an almost quaternionic structure Q on certain homogeneous manifolds $M = G/K$ (G is a Lie group and K a closed subgroup). The almost quaternionic structure Q will be G-invariant by construction and hence (M, Q) will be a homogeneous almost quaternionic manifold. Similarly, we will construct homogeneous almost quaternionic (pseudo-) Hermitian manifolds $(M = G/K, Q, g)$. In order to prove that $(M = G/K, Q, g)$ is a quaternionic (pseudo-) Kähler manifold, or simply to establish the 1-integrability of the almost quaternionic structure Q it is useful to have an appropriate description of the affine space of G-invariant connections on the homogeneous manifold $M = G/K$. This is provided by the notion of Nomizu map, which we now recall, see [**A-V-L**], [**V1**]. Let ∇ be a connection on a manifold M. For any vector field X on M one defines the operator

$$L_X := \mathcal{L}_X - \nabla_X\,, \tag{8}$$

where $\mathcal{L}_X$ is the Lie derivative (i.e. $\mathcal{L}_X Y = [X, Y]$ for any vector field Yon M). L_X is a $C^\infty(M)$-linear map on the $C^\infty(M)$-module $\Gamma(TM)$ of vector fields on M, so it can be identified with a section $L_X \in \Gamma(\mathrm{End}(TM))$. In particular, $L_X|_m \in \mathrm{End}(T_m M)$ for all $m \in M$.

Now let $M = G/K$ be a homogeneous space and suppose that ∇ is G-invariant. The action of G on M defines an antihomomorphism of Lie algebras $\alpha : \mathfrak{g} = \mathrm{Lie}\, G \to \Gamma(TM)$ from the Lie algebra $\mathfrak{g}$ of *left*-invariant vector fields on G to the Lie algebra $\Gamma(TM)$ of vector fields on M. This antihomomorphism maps an element $x \in \mathfrak{g}$ to the **fundamental vector field** $\alpha(x) := X \in \Gamma(TM)$ defined by $X(m) := \frac{d}{dt}|_{t=0} \exp(tx)m$. The statement that α is an antihomomorphism means that $[\alpha(x), \alpha(y)] = -\alpha([x, y])$ for all $x, y \in \mathfrak{g}$. Note that α becomes a homomorphism if we replace $\mathfrak{g}$ by the Lie algebra of *right*-invariant vector fields on G. Without restriction of generality, we can assume that the action is almost effective, i.e. $\mathfrak{g} \to \Gamma(TM)$ is injective. Then we can identify $\mathfrak{g}$ with its faithful image in $\Gamma(TM)$. The isotropy subalgebra $\mathfrak{k} = \mathrm{Lie}\, K$ is mapped (anti)isomorphically onto a subalgebra of vector fields vanishing at the base point $[e] = eK \in M = G/K$. In this situation we define the **Nomizu map** $L = L(\nabla) : \mathfrak{g} \to \mathrm{End}(T_{[e]}M)$, $x \mapsto L_x$, by the equation

$$L_x := L_X|_{[e]}\,,$$

where again X is the fundamental vector field on M associated to $x \in \mathfrak{g}$. The operators $L_x \in \mathrm{End}(T_{[e]}M)$ will be called **Nomizu operators**. They have the following properties:

$$L_x = d\rho(x) \quad \text{for all} \quad x \in \mathfrak{k} \tag{9}$$

and

$$L_{\mathrm{Ad}_k x} = \rho(k) L_x \rho(k)^{-1} \quad \text{for all} \quad x \in \mathfrak{g}\,, \quad k \in K\,, \tag{10}$$

where $\rho : K \to \mathrm{GL}(T_{[e]}M)$ is the isotropy representation. The first equation follows directly from equation (8) since $(\nabla_X Y)_m = 0$ if $X(m) = 0$. The second equation expresses the G-invariance of ∇. Conversely, any linear map $L : \mathfrak{g} \to \mathrm{End}(T_{[e]}M)$ satisfying (9) and (10) is the Nomizu map of a uniquely defined G-invariant connection $\nabla = \nabla(L)$ on M. Its torsion tensor T and curvature tensor R are expressed at $[e]$ by:

$$T(\pi x, \pi y) = -(L_x \pi y - L_y \pi x + \pi[x,y])$$

and

$$R(\pi x, \pi y) = [L_x, L_y] + L_{[x,y]}\,, \quad x, y \in \mathfrak{g}$$

where $\pi : \mathfrak{g} \to T_{[e]}M$ is the canonical projection $x \mapsto \pi x = X([e]) = \frac{d}{dt}|_{t=0} \exp(tx)K$.
Remark 6: The difference between our formulas for torsion and curvature and those of [**A-V-L**] is due to the fact that in [**A-V-L**] everything is expressed in terms of right-invariant vector fields on the Lie group G whereas we use left-invariant vector fields. To obtain the corresponding expressions in terms of right-invariant vector fields from the expressions in terms of left-invariant vector fields and vice versa it is sufficient to replace $[x,y]$ by $-[x,y]$ in all formulas $(x, y \in \mathfrak{g})$. The same remark applies to formula (11) below, which expresses the Levi-Civita connection on a pseudo-Riemannian homogeneous manifold.

Suppose now that we are given a G-invariant geometric structure S on M (e.g. a G-invariant almost quaternionic structure Q) defined by a corresponding K-invariant geometric structure $S_{[e]}$ on $T_{[e]}M$. Then a G-invariant connection ∇ preserves S if and only if the corresponding Nomizu operators L_x, $x \in \mathfrak{g}$, preserve $S_{[e]}$. So to construct a G-invariant connection preserving S it is sufficient to find a Nomizu map $L : \mathfrak{g} \to \mathrm{End}(T_{[e]}M)$ such that L_x preserves $S_{[e]}$ for all $x \in \mathfrak{g}$. We observe that, due to the K-invariance of $S_{[e]}$, the Nomizu operators L_x preserve $S_{[e]}$ already for $x \in \mathfrak{k}$. The above considerations can be specialized as follows:

PROPOSITION 8. *Let Q be a G-invariant almost quaternionic structure on a homogeneous manifold $M = G/K$. There is a natural one-to-one correspondence between G-invariant almost quaternionic connections on (M,Q) and Nomizu maps $L : \mathfrak{g} \to \mathrm{End}(T_{[e]}M)$, whose image normalizes $Q_{[e]}$, i.e. whose Nomizu operators L_x, $x \in \mathfrak{g}$, belong to the normalizer $\mathrm{n}(Q) \cong \mathfrak{sp}(1) \oplus \mathfrak{gl}(d, \mathbb{H})$ $(d = \dim M/4)$ of the quaternionic structure $Q_{[e]}$ in the Lie algebra $\mathfrak{gl}(T_{[e]}M)$.*

COROLLARY 4. *Let $(M = G/K, Q)$ be a homogeneous almost quaternionic manifold and $L : \mathfrak{g} \to \mathrm{End}(T_{[e]}M)$ a Nomizu map such that*

(1) *$L_x\pi y - L_y\pi x = -\pi[x, y]$ for all $x, y \in \mathfrak{g}$ (i.e. $T = 0$) and*
(2) *L_x normalizes $Q_{[e]} \subset \mathrm{End}(T_{[e]}M)$.*

Then the connection $\nabla(L)$ associated to the Nomizu map L is a G-invariant quaternionic connection on (M, Q) and hence Q is 1-integrable.

For future use we give the well known formula for the Nomizu map L^g associated to the Levi-Civita connection ∇^g of a G-invariant pseudo-Riemannian metric g on a homogeneous space $M = G/K$. Let $\langle \cdot, \cdot \rangle = g_{[e]}$ be the K-invariant scalar product on $T_{[e]}M$ induced by g. Then $L_x^g \in \mathrm{End}(T_{[e]}M)$, $x \in \mathfrak{g}$, is given by the following Koszul type formula:

$$-2\langle L_x^g \pi y, \pi z\rangle = \langle \pi[x, y], \pi z\rangle - \langle \pi x, \pi[y, z]\rangle - \langle \pi y, \pi[x, z]\rangle\,, \quad x, y, z \in \mathfrak{g}\,. \tag{11}$$

COROLLARY 5. *Let $(M = G/K, Q, g)$ be a homogeneous almost quaternionic (pseudo-) Hermitian manifold and assume that L_x^g normalizes $Q_{[e]}$ for all $x \in \mathfrak{g}$. Then the Levi-Civita connection $\nabla^g = \nabla(L^g)$ is a G invariant quaternionic connection on (M, Q, g) and hence (M, Q, g) is a quaternionic (pseudo-) Kähler manifold if* $\dim M > 4$.

2.4. The main theorem. Let $\mathfrak{p}(\Pi) = \mathfrak{p}(V) + W$ be an extended Poincaré algebra of signature (p, q) and $p \geq 3$. We fix the decomposition $p = p' + p''$, where $p' = 3$. For notational convenience we put $r := p'' = p - 3$. Then we consider the linear groups $K = K(p', p'') = K(3, r) \subset G(V) \subset G = G(\Pi) \subset \mathrm{Aut}(\mathfrak{r})$ introduced in section 2.1. As before $\mathfrak{r}$ denotes the radical of $\mathfrak{g} = \mathrm{Lie}\, G$.

THEOREM 8. 1) *There exists a G-invariant quaternionic structure Q on $M = M(\Pi) = G(\Pi)/K$.*

2) *If Π is nondegenerate (see Def. 2) then there exists a G-invariant pseudo-Riemannian metric g on M such that (M, Q, g) is a quaternionic pseudo-Kähler manifold.*

Proof: The main idea of the proof, which we will carry out in detail, is first to observe that the submanifold $M(V) = G(V)/K \subset G/K = M$ has a natural $G(V)$-invariant structure of quaternionic pseudo-Kähler manifold and then to study the problem of extending this structure to the manifold M. As shown in section 2.1, we can embed $M(V)$ as open $G(V)$-orbit into the pseudo-Riemannian symmetric space $\tilde{M}(V) = \mathrm{SO}_0(\tilde{V})/\tilde{K}$, $\tilde{V} = \mathbb{R}^{p+1,q+1}$, $\tilde{K} = \mathrm{SO}(p'+1) \times \mathrm{SO}_0(p'', q+1) = \mathrm{SO}(4) \times \mathrm{SO}_0(r, q+1)$. Now we claim that $\tilde{M}(V)$ carries an $\mathrm{SO}_0(\tilde{V})$-invariant (and hence $G(V)$-invariant) almost quaternionic structure Q. It is sufficient to specify the corresponding $\tilde{K}$-invariant quaternionic structure $Q_{[e]} \subset \mathrm{End}(T_{[e]}\tilde{M}(V))$ at the canonical base point $[e] = e\tilde{K} \in \tilde{M}(V)$. We first define a hypercomplex structure (J_α) on the isotropy module $T_{[e]}\tilde{M}(V) \cong \mathbb{R}^4 \otimes \mathbb{R}^{r,q+1}$:

$$J_1 := \mu_l(i) \otimes \mathrm{Id}\,, \quad J_2 := \mu_l(j) \otimes \mathrm{Id}\,, \quad J_3 := \mu_l(k) \otimes \mathrm{Id}\,. \tag{12}$$

Here $\mu_l(x) \in \mathrm{End}(\mathbb{R}^4)$ stands for left-multiplication by the quaternion $x \in \mathbb{H} = \mathrm{span}\{1, i, j, k\}$: $\mu_l(x)y = xy$ for all $x, y \in \mathbb{H} = \mathbb{R}^4$. Right-multiplication by x will be denoted by $\mu_r(x)$. Now it is easy to check that the quaternionic structure $Q_{[e]} := \mathrm{span}\{J_1, J_2, J_3\}$ is normalized under the isotropy representation of $\tilde{K}$ and hence extends to an $\mathrm{SO}_0(\tilde{V})$-invariant almost quaternionic structure Q on $\tilde{M}(V)$. Note that the complex structures J_α are not invariant under the isotropy representation and hence do not extend to $SO_0(\tilde{V})$-invariant almost complex structures on $\tilde{M}(V)$. Next we observe that the tensor product of the canonical pseudo-Euclidean scalar products on $\mathbb{R}^4$ and $\mathbb{R}^{r,q+1}$ defines a pseudo-Euclidean scalar product $-g_{[e]}$ on $T_{[e]}\tilde{M}(V) \cong \mathbb{R}^4 \otimes \mathbb{R}^{r,q+1}$. Notice that $g_{[e]}$ is *positive* definite if $r = 0$ and indefinite otherwise. It is invariant under the isotropy representation and hence extends to a (Q-Hermitian) $SO_0(\tilde{V})$-invariant pseudo-Riemannian metric on $\tilde{M}(V)$, which is unique, up to scaling, and defines on $\tilde{M}(V)$ the well known structure of pseudo-Riemannian symmetric space. In fact, it is the symmetric space associated to the symmetric pair $(\mathfrak{o}(\tilde{V}), \tilde{\mathfrak{k}} := \mathrm{Lie}\,\tilde{K})$. Let $\tilde{\mathfrak{m}} \subset \mathfrak{o}(\tilde{V})$ be the $\tilde{\mathfrak{k}}$-invariant complement to the isotropy algebra $\tilde{\mathfrak{k}}$. Then $[\tilde{\mathfrak{m}}, \tilde{\mathfrak{m}}] \subset \tilde{\mathfrak{k}}$ and hence the Nomizu operators $L^g_x \in \mathrm{End}(T_{[e]}\tilde{M}(V))$ associated to the Levi-Civita connection ∇^g of g vanish for $x \in \tilde{\mathfrak{m}}$, see (11). On the other hand if $x \in \tilde{\mathfrak{k}}$ then the Nomizu operator L^g_x on $T_{[e]}\tilde{M}(V) \cong \tilde{\mathfrak{m}}$ coincides with the image of x under the isotropy representation: $L^g_x = \mathrm{ad}_x|\tilde{\mathfrak{m}}$. This shows that L^g_x normalizes the quaternionic structure $Q_{[e]}$ for all $x \in \mathfrak{o}(\tilde{V})$ and by Cor. 5 we conclude that $(\tilde{M}(V), Q, g)$ is a homogeneous quaternionic pseudo-Kähler manifold if $\dim \tilde{M}(V) > 4$. The manifold $\tilde{M}(V)$ is 4-dimensional only if $r = q = 0$ and in this case $\tilde{M}(V) = SO_0(4,1)/SO(4)$ reduces to real hyperbolic 4-space $H^4_{\mathbb{R}}$, i.e. to the quaternionic hyperbolic line $H^1_{\mathbb{H}} = Sp(1,1)/Sp(1) \cdot Sp(1)$, which is a standard example of (conformally half-flat Einstein) quaternionic Kähler 4-manifold. Of course, the pair (Q, g) defined on the manifold $\tilde{M}(V)$ (for all r and q) restricts to a $G(V)$-invariant quaternionic pseudo-Kähler structure on the open $G(V)$-orbit $M(V) \hookrightarrow \tilde{M}(V)$. So we have proven that $(M(V) = G(V)/K, Q, g)$ is a homogeneous quaternionic pseudo-Kähler manifold.

Our strategy is now to extend the geometric structures Q and g from $M(V)$ to G-invariant structures on $M = G/K \supset G(V)/K = M(V)$. First we will extend Q to a G-invariant almost quaternionic structure on M. Using the infinitesimal action of $\mathfrak{g} = \mathfrak{g}(\Pi) = \mathfrak{g}(V) + W$ on M we can identify $T_{[e]}M = (\mathfrak{g}(V)/\mathfrak{k}) \oplus W = T_{[e]}M(V) \oplus W$. Note that the isotropy representation of K on $T_{[e]}M$ preserves this decomposition and acts on $T_{[e]}M(V) = T_{eK}M(V) \cong T_{e\tilde{K}}\tilde{M}(V)$ as restriction of the isotropy representation of $\tilde{K}$ on $T_{e\tilde{K}}\tilde{M}(V)$ to the subgroup $K \subset \tilde{K}$. We extend the hypercomplex structure (J_α) defined above on $T_{e\tilde{K}}\tilde{M}(V) \cong T_{eK}M(V)$ to a hypercomplex structure on $T_{eK}M = T_{eK}M(V) \oplus W$ as follows:

$$J_\alpha s := e_\beta e_\gamma s\,, \quad s \in W\,, \tag{13}$$

where $(e_\alpha, e_\beta, e_\gamma)$ is a cyclic permutation of the standard orthonormal basis of $E = \mathrm{R}^{3,0} \subset V = E + E'$ (the product $e_\beta e_\gamma$ is in the Clifford algebra). Now we can extend the quaternionic structure $Q_{[e]}$ on $T_{[e]}M(V)$ to a quaternionic structure on $T_{[e]}M$ such that $Q_{[e]} = \mathrm{span}\{J_1, J_2, J_3\}$. To prove that $Q_{[e]} \subset \mathrm{End}(T_{[e]}M)$ extends to a G-invariant almost quaternionic structure on M it is sufficient to check that the isotropy algebra $\mathfrak{k}$ normalizes $Q_{[e]}$. It is obvious that the subalgebra $\mathfrak{o}(E') = \mathfrak{o}(r,q) \subset \mathfrak{k} = \mathfrak{o}(E) \oplus \mathfrak{o}(E')$ centralizes $Q_{[e]}$. We have to show that $\mathfrak{o}(E) = \mathfrak{o}(3)$ normalizes $Q_{[e]}$. In terms of the standard basis $(e_1, e_2, e_3) = (i, j, k)$ of $E = \mathbb{R}^3 = \mathrm{Im}\mathbb{H}$ a basis of $\mathfrak{o}(3) = \wedge^2 E$ is given by $(e_2 \wedge e_3, e_3 \wedge e_1, e_1 \wedge e_2)$. For any cyclic permutation (α, β, γ) of $(1,2,3)$ we have the following easy formulas, cf. (4):

$$-2d\rho(e_\beta \wedge e_\gamma) = (\mu_l(e_\alpha) - \mu_r(e_\alpha)) \otimes \mathrm{Id} \tag{14}$$

on $T_{eK}M(V) \cong T_{e\tilde{K}}\tilde{M}(V) \cong \mathbb{R}^4 \otimes \mathbb{R}^{r,q+1}$ and

$$-2d\rho(e_\beta \wedge e_\gamma) = e_\beta e_\gamma = J_\alpha \tag{15}$$

on W. As before $d\rho : \mathfrak{k} \to \mathfrak{gl}(T_{[e]}M)$ denotes the isotropy representation. From the equations (12)-(15) we immediately obtain that

$$[d\rho(e_\beta \wedge e_\gamma), J_\alpha] = 0 \quad \text{and} \quad [d\rho(e_\beta \wedge e_\gamma), J_\beta] = -J_\gamma .$$

This shows that $Q_{[e]} \subset \mathrm{End}(T_{[e]}M)$ is invariant under $\mathfrak{k}$ and hence extends to a G-invariant almost quaternionic structure Q on M. The 1-integrability of Q will be proven in the sequel.

Let us first treat the case where Π is nondegenerate. First of all, we extend the $G(V)$-invariant Q-Hermitian pseudo-Riemannian metric g on $M(V)$ to a G-invariant Q-Hermitian pseudo-Riemannian metric on M. It is sufficient to extend the K-invariant pseudo-Euclidean scalar product $g_{[e]}$ on $T_{[e]}M(V)$ to a K-invariant and J_α-invariant ($\alpha = 1, 2, 3$) scalar product on $T_{[e]} = T_{[e]}M(V) \oplus W$. We do this in such a way that the above decomposition is orthogonal for the extended scalar product $g_{[e]}$ on $T_{[e]}M$ and define

$$g_{[e]}|W \times W := -b ,$$

where $b = b_{\Pi,(e_1,e_2,e_3)}$ is the canonical symmetric bilinear form associated to Π and to the decomposition $p = p' + p'' = 3 + r$. By Prop. 1 the nondegeneracy of Π implies the nondegeneracy of b. This shows that the symmetric bilinear form $g_{[e]}$ on $T_{[e]}M$ defined above is indeed a pseudo-Euclidean scalar product. This scalar product is invariant under the isotropy group $K = K(p', p'') = K(3, r)$, by virtue of Thm. 1, 3), and hence extends to a G-invariant pseudo-Riemannian metric g on M. Moreover, the metric g is Q-Hermitian. In fact, it is sufficient to observe that b is J_α-invariant. This property follows from the K-invariance of b, since $J_\alpha|W \in d\rho(\mathfrak{k})|W$, see (15). Summarizing, we obtain: (M, Q, g) is a homogeneous pseudo-Hermitian almost quaternionic manifold. The next step is to compute the Nomizu map $L^g : \mathfrak{g} \to \mathrm{End}(T_{[e]}M)$ associated to the Levi-Civita connection ∇^g of g. It is convenient to identify $T_{[e]}M$ with a $\mathfrak{k}$-invariant complement $\mathfrak{m}$ to $\mathfrak{k}$ in $\mathfrak{g}$. Such complement is easily described in terms of the K-invariant orthogonal decomposition $V = E + E' = \mathbb{R}^{3,0} + \mathbb{R}^{r,q}$. Indeed, using the canonical identification $\mathfrak{o}(V) = \wedge^2 V$ we have the following $\mathfrak{k}$-invariant decompositions:

$$\mathfrak{o}(V) = \mathfrak{k} + E \wedge E' ,$$

$$\mathfrak{g}(V) = \mathfrak{o}(V) + \mathbb{R}D + V = \mathfrak{k} + \mathfrak{m}(V) , \quad \mathfrak{m}(V) = E \wedge E' + \mathbb{R}D + V ,$$

$$\mathfrak{g} = \mathfrak{g}(V) + W = \mathfrak{k} + \mathfrak{m} , \quad \mathfrak{m} = \mathfrak{m}(\Pi) := \mathfrak{m}(V) + W .$$

In the following, we will make the identifications $T_{[e]}M(V) = \mathfrak{g}(V)/\mathfrak{k} = \mathfrak{m}(V) \subset T_{[e]}M = \mathfrak{g}/\mathfrak{k} = \mathfrak{m}$. Let (e'_a), $a = 1, \ldots, r + q$, be the standard orthonormal basis of $E' = \mathbb{R}^{r,q}$. Then we have the following orthonormal basis of $(\mathfrak{m}(V), g_{[e]})$:

$$(e_\alpha \wedge e'_a, D, e_\alpha, e'_a) ,$$

where $\alpha = 1, 2, 3$ and $a = 1, \dots, r+q$. For notational convenience we denote this basis by $(e_{\iota i})$, $\iota = 0, \dots 3$, $i = 0, \dots, r+q$, where

$$\begin{aligned} e_{00} &:= e'_0 := e_0 := D \\ e_{0a} &:= e'_a \quad (a = 1, \dots, r+q) \\ e_{\alpha 0} &:= e_\alpha \quad (\alpha = 1, 2, 3) \\ e_{\alpha a} &:= e_\alpha \wedge e'_a \,. \end{aligned}$$

In terms of this basis the hypercomplex structure J_α is expressed on $\mathfrak{m}(V)$ simply by:

$$J_\alpha e_{0i} = J_\alpha e'_i = e_{\alpha i} \quad \text{and} \quad J_\alpha e_{\beta i} = e_{\gamma i}\,,$$

where $i = 0, \dots, r+q$ and (α, β, γ) is a cyclic permutation of $(1, 2, 3)$. The scalar product $g_{[e]}$ is completely determined on $\mathfrak{m}(V)$ by the condition that $(e_{\iota i})$ is orthonormal and that

$$\epsilon_i := g(e_{\iota i}, e_{\iota i}) = \begin{cases} -1 & \text{if} \quad i = 1, \dots, r \\ +1 & \text{if} \quad i = 0, r+1, \dots, r+q\,. \end{cases}$$

The scalar product $g_{[e]}$ on $\mathfrak{m}$ induces an identification $x \wedge y \mapsto x \wedge_g y$ of the exterior square $\wedge^2 \mathfrak{m}$ with the vector space of $g_{[e]}$-skew symmetric endomorphisms on $\mathfrak{m}$, where $x \wedge_g y(z) := g(y, z)x - g(x, z)y$, $x, y, z \in \mathfrak{m}$. We denote by $\mathrm{n}(Q)$ (respectively, $\mathrm{z}(Q)$) the normalizer (respectively, centralizer) of the quaternionic structure $Q_{[e]}$ in the Lie algebra $\mathfrak{gl}(\mathfrak{m})$.

LEMMA 5. *The Nomizu map $L^g = L(\nabla^g)$ associated to the Levi-Civita connection ∇^g of the homogeneous pseudo-Riemannian manifold $(M = G/K, g)$ is given by the following formulas:*

$$\begin{aligned} L^g_{e_0} &= L^g_{e_{\alpha a}} = 0\,, \\ L^g_{e_\alpha} &= \frac{1}{2} J_\alpha + \bar{L}^g_{e_\alpha}\,, \end{aligned}$$

where

$$\bar{L}^g_{e_\alpha} = \frac{1}{2} \sum_{i=0}^{r+q} \epsilon_i e_{\alpha i} \wedge_g e'_i - \frac{1}{2} \sum_{i=0}^{r+q} \epsilon_i e_{\gamma i} \wedge_g e_{\beta i} \in \mathrm{z}(Q)\,,$$

$L^g_{e'_a} \in \mathrm{z}(Q)$ is given by:

$$\begin{aligned} L^g_{e'_a} | \mathfrak{m}(V) &= \sum_{\iota=0}^{3} e_{\iota a} \wedge_g e_\iota\,, \\ L^g_{e'_a} | W &= \frac{1}{2} e_1 e_2 e_3 e'_a\,. \end{aligned}$$

For all $s \in W$ the Nomizu operator $L^g_s \in \mathrm{z}(Q)$ maps the subspace $\mathfrak{m}(V) \subset \mathfrak{m} = \mathfrak{m}(V) + W$ into W and W into $\mathfrak{m}(V)$. The restriction $L^g_s | W$ $(s \in W)$ is completely determined by $L^g_s | \mathfrak{m}(V)$ (and vice versa) according to the relation

$$g(L^g_s t, x) = -g(t, L^g_s x)\,, \quad s, t \in W\,, \quad x \in \mathfrak{m}(V)\,.$$

Finally, $L^g_s|\mathfrak{m}(V)$ ($s \in W$) is completely determined by its values on the quaternionic basis (e'_i), $i = 0, \ldots, r+q$, which are as follows:

$$\begin{aligned} L^g_s e'_0 &= \frac{1}{2} s\,, \\ L^g_s e'_a &= \frac{1}{2} e_1 e_2 e_3 e'_a s\,. \end{aligned}$$

(It is understood that $L^g_x = d\rho(x) = \mathrm{ad}_x|\mathfrak{m}$ for all $x \in \mathfrak{k}$, cf. equation (9).) In the above formulas $a = 1, \ldots, r+q$, $\alpha = 1, 2, 3$ and (α, β, γ) is a cyclic permutation of $(1, 2, 3)$.

Proof: This follows from equation (11) by a straightforward computation. □

COROLLARY 6. *The Levi-Civita connection ∇^g of the homogeneous almost quaternionic pseudo-Hermitian manifold (M, Q, g) preserves Q and hence (M, Q, g) is a quaternionic pseudo-Kähler manifold.*

Proof: From the G-invariance of Q, we know already that $L^g_x \in \mathrm{n}(Q)$ for all $x \in \mathfrak{k}$ and the formulas of Lemma 5 show that $L^g_x \in \mathrm{n}(Q)$ for all $x \in \mathfrak{m}$. Now the corollary follows from Cor. 5. □

By Cor. 6 we have already established part 2) of Thm. 8. Part 1) is a consequence of 2) provided that Π is nondegenerate. It remains to discuss the case of degenerate Π.

By Thm. 2, we have a direct decomposition $W = W_0 + W'$ of $C\ell^0(V)$-modules, where $W_0 = \ker \Pi$. We put $\Pi' := \Pi|\wedge^2 W'$ and denote by $L' : \mathfrak{g}(\Pi') \to \mathrm{End}(\mathfrak{m}(\Pi'))$ the Nomizu map associated to the Levi-Civita connection of the quaternionic pseudo-Kähler manifold $(M' := M(\Pi'), Q, g)$. Note that the quaternionic structure Q of M' coincides with the almost quaternionic structure induced by the obvious $G(\Pi')$-equivariant embedding $M' = G(\Pi')/K \subset G(\Pi)/K = M$. By the next lemma, we can extend the map L' to a torsionfree Nomizu map $L : \mathfrak{g}(\Pi) \to \mathrm{End}(\mathfrak{m}(\Pi))$, whose image normalizes $Q_{[e]}$. This proves the 1-integrability of Q (by Cor. 4), completing the proof of Thm. 8. □

By Cor. 6 we can decompose $L'_x = \sum_{\alpha=1}^3 \omega'_\alpha(x) J_\alpha + \bar{L}'_x$, where the ω'_α are 1-forms on $\mathfrak{m}(\Pi')$ and $\bar{L}'_x \in \mathrm{z}(Q)$ belongs to the centralizer of the quaternionic structure on $\mathfrak{m}(\Pi')$.

LEMMA 6. *The Nomizu map $L' : \mathfrak{g}(\Pi') \to \mathrm{End}(\mathfrak{m}(\Pi'))$ associated to the Levi-Civita connection of $(M(\Pi') = G(\Pi')/K, g)$ can be extended to the Nomizu map $L : \mathfrak{g}(\Pi) \to \mathrm{End}(\mathfrak{m}(\Pi))$ of a $G(\Pi)$-invariant quaternionic connection ∇ on the homogeneous almost quaternionic manifold $(M(\Pi) = G(\Pi)/K, Q)$. The extension is defined as follows:*

$$L_x := \sum_{\alpha=1}^{3} \omega_\alpha(x) J_\alpha + \bar{L}_x\,, \quad x \in \mathfrak{m}(\Pi)\,,$$

where $\bar{L}_x \in \mathrm{z}(Q)$, with centralizer taken in $\mathfrak{gl}(\mathfrak{m}(\Pi))$, is defined below and the 1-forms ω_α on $\mathfrak{m}(\Pi) := \mathfrak{m}(\Pi') + W_0$ satisfy $\omega_\alpha|\mathfrak{m}(\Pi') := \omega'_\alpha$ and $\omega_\alpha|W_0 := 0$. The operators $\bar{L}_x$ are given by:

$$\bar{L}_x|\mathfrak{m}(\Pi') := \bar{L}'_x \quad \text{if} \quad x \in \mathfrak{m}(\Pi')\,,$$

$$\bar{L}_{e_0}|W_0 := \bar{L}_{e_{\alpha a}}|W_0 := \bar{L}_{e_\alpha}|W_0 := 0\,,$$

$$\bar{L}_{e'_a}|W_0 := \frac{1}{2} e_1 e_2 e_3 e'_a ,$$

$$\bar{L}_s|W_0 := 0 \quad \textit{if} \quad s \in W ,$$

$$\bar{L}_s x := L_x s - [s, x] \quad \textit{if} \quad s \in W_0 , \quad x \in \mathfrak{m}(\Pi') .$$

(It is understood that $L_x = \mathrm{ad}_x|\mathfrak{m}(\Pi)$ for all $x \in \mathfrak{k}$. In the above formulas, as usual, $a = 1, \ldots, r+q$ and $\alpha = 1, 2, 3$.)

Proof: To check that $L : \mathfrak{g}(\Pi) \to \mathrm{End}(\mathfrak{m}(\Pi))$ is a Nomizu map it is sufficient to check (10); (9) is satisfied by definition. The equation (10) expresses the K-invariance of $L \in \mathfrak{g}(\Pi)^* \otimes \mathrm{End}(\mathfrak{m}(\Pi))$ which is equivalent to the invariance of L under the Lie algebra $\mathfrak{k}$, since K is a connected Lie group. So (10) is equivalent to the following equation:

$$L_{[x,y]} z = [\mathrm{ad}_x, L_y] z \quad x \in \mathfrak{k}, \quad y, z \in \mathfrak{m}(\Pi) . \tag{16}$$

We first check this equation. Let always $x \in \mathfrak{k}$. Due to $L_W W_0 = 0$ and $[\mathfrak{k}, W_0] \subset W_0$ we have

$$L_{[x,y]} z = [\mathrm{ad}_x, L_y] z = 0$$

if $y \in W$ and $z \in W_0$. Also (16) is satisfied if $y, z \in \mathfrak{m}(\Pi')$ because $L_x|\mathfrak{m}(\Pi') = L'_x$ is the Nomizu operator associated to the Nomizu map L'. Now let $y \in W_0$ and $z \in \mathfrak{m}(\Pi')$. Then we compute:

$$\begin{aligned} L_{[x,y]} z - [\mathrm{ad}_x, L_y] z &= (L_z[x,y] - [[x,y],z]) - [x, L_y z] + L_y[x,z] \\ &= L_z[x,y] - [[x,y],z] - [x, L_z y] + [x,[y,z]] + L_{[x,z]} y - [y,[x,z]] \\ &= L_{[x,z]} y - [x, L_z y] + L_z[x,y] = L_{[x,z]} y - [\mathrm{ad}_x, L_z] y . \end{aligned}$$

This shows that it is sufficient to check (16) for $y \in \mathfrak{m}(\Pi')$ and $z \in W_0$; the case $y \in W_0$ and $z \in \mathfrak{m}(\Pi')$ then follows from the above computation. In the following let $x \in \mathfrak{k}$ and $z \in W_0$. We check (16) for all $y \in \mathfrak{m}(\Pi')$. From $L_{e_0} = 0$ and $[\mathfrak{k}, e_0] = 0$ it follows that

$$L_{[x,e_0]} z - [x, L_{e_0} z] + L_{e_0}[x,z] = 0 .$$

Next we check (16) for $y = e_\alpha$:

$$L_{[x,e_\alpha]} z - [x, L_{e_\alpha} z] + L_{e_\alpha}[x,z] = L_{[x,e_\alpha]} z - \frac{1}{2}[x, J_\alpha z] + \frac{1}{2} J_\alpha [x,z] .$$

It is clear that the first summand and the sum of the second and third summands vanish if $x \in \mathfrak{o}(E') \subset \mathfrak{k} = \mathfrak{o}(E) \oplus \mathfrak{o}(E')$. For $x = e_\alpha \wedge e_\beta \in \mathfrak{o}(E) = \mathfrak{o}(3)$ ((α, β, γ) cyclic) we compute:

$$\begin{aligned} &L_{[e_\alpha \wedge e_\beta, e_\alpha]} z - \frac{1}{2}[e_\alpha \wedge e_\beta, J_\alpha z] + \frac{1}{2} J_\alpha [e_\alpha \wedge e_\beta, z] \\ &= -L_{e_\beta} z + \frac{1}{4} e_\alpha e_\beta e_\beta e_\gamma z - \frac{1}{4} e_\beta e_\gamma e_\alpha e_\beta z \\ &= -\frac{1}{2} e_\gamma e_\alpha z + \frac{1}{2} e_\gamma e_\alpha z = 0 \end{aligned}$$

and for $x = e_\beta \wedge e_\gamma$ we obtain:

$$L_{[e_\beta \wedge e_\gamma, e_\alpha]} z - \frac{1}{2}[e_\beta \wedge e_\gamma, J_\alpha z] + \frac{1}{2} J_\alpha [e_\beta \wedge e_\gamma, z]$$

$$= 0 + \frac{1}{4} e_\beta e_\gamma e_\beta e_\gamma z - \frac{1}{4} e_\beta e_\gamma e_\beta e_\gamma z = 0 .$$

Next we check (16) for $y = e'_a$:

$$L_{[x,e'_a]} z - [x, L_{e'_a} z] + L_{e'_a}[x, z] =$$

$$\frac{1}{2} e_1 e_2 e_3 [x, e'_a] z - \frac{1}{2}[x, e_1 e_2 e_3 e'_a z] + \frac{1}{2} e_1 e_2 e_3 e'_a [x, z] .$$

It is easy to see that this is zero if $[x, e'_a] = 0$. So we can put $x = e'_b \wedge e'_a$ obtaining:

$$\frac{1}{2}\langle e'_a, e'_a \rangle e_1 e_2 e_3 e'_b z + \frac{1}{4} e'_b e'_a e_1 e_2 e_3 e'_a z - \frac{1}{4} e_1 e_2 e_3 e'_a e'_b e'_a z = 0 .$$

Finally, for $y \in E \wedge E'$ we immediately obtain:

$$L_{[x,y]} z - [x, L_y z] + L_y [x, z] = L_{[x,y]} z = 0 ,$$

since $L_y = 0$ for $y \in E \wedge E'$ and $[\mathfrak{k}, E \wedge E'] \subset E \wedge E'$. So we have proven that L is a Nomizu map. It is easily checked that $L_x y - L_y x = -\pi[x, y]$ for all $x, y \in \mathfrak{m}(\Pi)$, where $\pi : \mathfrak{g}(\Pi) \to \mathfrak{m}(\Pi) \cong T_{[e]} M(\Pi)$ is the projection along $\mathfrak{k}$. This shows that $\nabla = \nabla(L)$ has zero torsion. It only remains to check that $L_x \in \mathrm{n}(Q)$ for all $x \in \mathfrak{g}(\Pi)$. This is easy to see for $x \in \mathfrak{g}(\Pi')$ from the definition of the map L as extension of the Nomizu map L'. We present the calculation only for L_s, $s \in W_0$:

$$L_s e_0 = L_{e_0} s - [s, e_0] = 0 + \frac{1}{2} s = \frac{1}{2} s ,$$

$$L_s J_\alpha e_0 = L_s e_\alpha = L_{e_\alpha} s - [s, e_\alpha] = \frac{1}{2} J_\alpha s - 0 = J_\alpha L_s e_0 ,$$

$$L_s e'_a = L_{e'_a} s - [s, e'_a] = \frac{1}{2} e_1 e_2 e_3 e'_a s - 0 = \frac{1}{2} e_1 e_2 e_3 e'_a s ,$$

$$L_s J_\alpha e'_a = L_s e_{\alpha a} = L_{e_{\alpha a}} s - [s, e_{\alpha a}] = 0 - \frac{1}{2} e_\alpha e'_a s$$

$$= e_\beta e_\gamma (\frac{1}{2} e_\alpha e_\beta e_\gamma e'_a s) = J_\alpha L_s e'_a ,$$

(α, β, γ) cyclic,

$$L_s t = L_t s - [s, t] = 0 ,$$

$$L_s J_\alpha t = L_{J_\alpha t} s - [s, J_\alpha t] = 0 = J_\alpha L_t s$$

if $t \in W$. We have used that $L_W W_0 = 0$ and $[W_0, W] = 0$. This shows that $L_s \in \mathrm{z}(Q)$ for all $s \in W_0$ finishing the proof of the lemma. □

2.5. The Riemannian case.

PROPOSITION 9. *Let $\mathfrak{p}(\Pi) = \mathfrak{p}(V) + W$ be a nondegenerate extended Poincaré algebra of signature (p, q), $p \geq 3$, and $(M(\Pi), Q, g)$ the corresponding homogeneous quaternionic pseudo-Kähler manifold, see Thm. 8. Then the pseudo-Riemannian metric g is positive definite, and hence a Riemannian metric, if and only if $-b$ is positive definite and $p = 3$. In all other cases g is indefinite. Here $b = b_{\Pi,(e_1,e_2,e_3)}$ is the canonical symmetric bilinear form associated to Π and to the decomposition $p = 3 + r$.*

Proof: By construction, the restriction of g to the submanifold $M(V) \subset M(\Pi)$ is a (positive definite) Riemannian metric if and only if $p = 3$ and is indefinite otherwise. Now the proposition follows from the fact that $-b$ is precisely the restriction of the scalar product $g_{[e]}$ on $T_{[e]}M(\Pi) \cong T_{[e]}M(V) \oplus W$ to the subspace W. □

Next we will use the classification of extended Poincaré algebras of signature $(3, q)$ up to isomorphism (see 1.3) to derive the classification of the quaternionic Kähler manifolds $(M(\Pi), Q, g)$ up to isometry. We recall that $\mathfrak{p}(3, q, 0, 0, l) = \mathfrak{p}(p = 3, q, l_0 = 0, l_+ = 0, l_- = l)$ (respectively, $\mathfrak{p}(3, q, 0, 0, l^+, 0, 0, l^-) = \mathfrak{p}(p = 3, q, l_0^+ = 0, l_+^+ = l^+, l_0^- = 0, l_+^- = 0, l_-^- = l^-)$) is the set of isomorphism classes of extended Poincaré algebras for which $-b$ is positive definite if $q \not\equiv 3 \pmod 4$ (respectively, if $q \equiv 3 \pmod 4$). We denote by $M(q, l)$ (respectively, $M(q, l^+, l^-)$) the homogeneous quaternionic Kähler manifold $(M(\Pi), Q, g)$ associated to $\Pi \in \mathfrak{p}(3, q, 0, 0, l)$ (respectively, $\Pi \in \mathfrak{p}(3, q, 0, 0, l^+, 0, 0, l^-)$).

THEOREM 9. *Every homogeneous quaternionic pseudo-Kähler manifold of the form $(M(\Pi), Q, g)$ for which g is positive definite is isometric to one of the homogeneous quaternionic Kähler manifolds $M(q, l)$ ($q \not\equiv 3 \pmod 4$) or $M(q, l^+, l^-) \cong M(q, l^-, l^+)$ ($q \equiv 3 \pmod 4$). In particular, there are only countably many such Riemannian manifolds up to isometry.*

Proof: This is a direct consequence of Prop. 9, Thm. 6 and Thm. 7. □

Any real vector space E admitting a quaternionic structure has $\dim E \equiv 0 \pmod 4$. Therefore, we can define its **quaternionic dimension** $\dim_{\mathbb{H}} E := \dim E/4$. Similarly, the quaternionic dimension of a quaternionic manifold (M, Q) is $\dim_{\mathbb{H}} M := \dim M/4$. We denote by $N(q)$ the quaternionic dimension of an irreducible $C\ell^0_{3,q}$-module.

PROPOSITION 10. 1) *$N(0) = N(1) = N(2) = N(3) = 1$, $N(4) = 2$, $N(5) = 4$, $N(6) = N(7) = 8$ and $N(q + 8) = 16N(q)$ for all $q \geq 0$. In particular, $N(q)$ coincides with the dimension of an irreducible $\mathbb{Z}_2$-graded $C\ell_{q-3}$-module if $q \geq 3$.*

2) *The quaternionic dimension of the homogeneous quaternionic Kähler manifolds $M(q, l)$ and $M(q, l^+, l^-)$ is given by:*

$$\dim_{\mathbb{H}} M(q, l) = q + 1 + lN(q)$$

and

$$\dim_{\mathbb{H}} M(q, l^+, l^-) = q + 1 + (l^+ + l^-)N(q)\,.$$

Proof: The first part follows from the classification of Clifford algebras. The second part follows from $\dim M = \dim M(V) + \dim W$. □

The next theorem identifies the spaces $M(q, l)$ and $M(q, l^+, l^-)$ with Alekseevsky's quaternionic Kähler manifolds. We recall that an **Alekseevsky space** is a quaternionic Kähler manifold which admits a simply transitive non Abelian splittable solvable group of isometries, see [**A3**] and [**C2**]. Due to Iwasawa's decomposition theorem any symmetric quaternionic Kähler manifold of noncompact type is an Alekseevsky space. These are precisely the noncompact duals of the Wolf spaces. We recall that a **Wolf space** is a symmetric quaternionic Kähler manifold of compact type and that such manifolds are in 1-1-correspondence with the complex simple Lie algebras [**W1**]. The nonsymmetric Alekseevsky spaces are grouped into 3 series: $\mathcal{V}$-spaces, $\mathcal{W}$-spaces and $\mathcal{T}$-spaces, see [**A3**], [**dW-VP2**] and [**C2**]. These 3 series contain also all symmetric Alekseevsky spaces of rank > 2

and no symmetric spaces of smaller rank. By definition an Alekseevsky space can be presented as metric Lie group, i.e. as homogeneous Riemannian manifold of the form (L, g), where L is a Lie group and g a left-invariant Riemannian metric on L.

THEOREM 10. *Let* $(M = M(\Pi) = G(\Pi)/K, Q, g)$ *be a homogeneous quaternionic Kähler manifold as in Prop. 9,*

$$G(\Pi) = S \ltimes R \cong \mathrm{Spin}_0(V) \ltimes R$$

the Levi decomposition (7) and $I(S)$ *the Iwasawa subgroup of* S. *Then* $L := L(\Pi) := I(S) \ltimes R \subset G(\Pi)$ *is a (non Abelian) splittable solvable Lie subgroup which acts simply transitively on* M. *In particular,* (M, Q, g) *is an Alekseevsky space. More precisely, we have the following identifications with the* $\mathcal{V}$*-spaces,* $\mathcal{W}$*-spaces,* $\mathcal{T}$*-spaces and symmetric Alekseevsky spaces:*

1) $M(q, l) = \mathcal{V}(l, q-3)$ *and* $M(q, l^+, l^-) = \mathcal{V}(l^+, l^-, q-3)$ *if* $q \geq 4$,
2) $M(3, l^+, l^-) = \mathcal{W}(l^+, l^-)$,
3) $M(2, l) = \mathcal{T}(l)$,
4) $M(1, l) = \mathrm{SU}(l+2, 2)/\mathrm{S}(\mathrm{U}(l+2) \times \mathrm{U}(2))$,
5) $M(0, l) = \mathrm{Sp}(l+1, 1)/\mathrm{Sp}(l+1)\mathrm{Sp}(1) = \mathbb{H}H^{l+1}$ *(quaternionic hyperbolic* $(l+1)$*-space).*

The above Riemannian manifolds $M(q, l)$ *and* $M(q, l^+, l^-) \cong M(q, l^-, l^+)$ *are pairwise nonisometric and exhaust all Alekseevsky spaces with the following two symmetric exceptions: the complex hyperbolic plane* $\mathbb{C}H^2 = \mathrm{SU}(1,2)/\mathrm{U}(2) =: M(1, -1)$ *and* $\mathrm{G}_2^{(2)}/\mathrm{SO}(4)$. *(Note that these two symmetric spaces have rank* ≤ 2 *and so do not belong to any of the 3 series* $\mathcal{V}$, $\mathcal{W}$ *and* $\mathcal{T}$ *of Alekseevsky spaces.)*

Proof: The fact that L acts simply transitively on M follows from Prop. 7. This shows that (M, Q, g) is an Alekseevsky space. The Riemannian metric g induces a left-invariant metric g_L on the Lie group L. To establish the identifications given in the theorem it is sufficient to check that (L, g_L) is isomorphic (as metric Lie group) to one of the metric Lie groups which occur in the classification of Alekseevsky spaces, see [**A3**] and [**C2**]. (The quaternionic structure can be reconstructed from the holonomy of the Levi-Civita connection, up to an automorphism of the full isometry group which preserves the isotropy group.) Finally, to prove that $M(q, l)$ and $M(q, l^+, l^-) \cong M(q, l^-, l^+)$ are pairwise nonisometric it is, by [**A4**], sufficient to check that the corresponding metric Lie groups (which occur in the classification of Alekseevsky spaces) are pairwise nonisomorphic. This was done in [**C2**]. □

2.6. A class of noncompact homogeneous quaternionic Hermitian manifolds with no transitive solvable group of isometries. There is a widely known conjecture by D.V. Alekseevsky which says that any noncompact homogeneous quaternionic Kähler manifold admits a transitive solvable group of isometries [**A3**]. The next theorem shows that this conjecture becomes false if we replace "Kähler" by "Hermitian".

THEOREM 11. *Let* $\mathfrak{p}(\Pi)$ *be any extended Poincaré algebra of signature* $(p, q) = (3+r, 0)$, $r \geq 0$, *and* $(M = G/K, Q)$ *the corresponding homogeneous quaternionic manifold, see Thm. 8, 1). Then there exists a* G*-invariant* Q*-Hermitian Riemannian metric* h *on* M. *Moreover, the noncompact homogeneous quaternionic Hermitian manifold* (M, Q, h) *does not admit any transitive solvable Lie group of isometries if* $r > 0$.

Proof: Since $K = K(3,r) \cong \mathrm{Spin}(3) \cdot \mathrm{Spin}(r)$ is compact, one can easily construct a K-invariant $Q_{[e]}$-Hermitian Euclidean scalar product $h_{[e]}$ on $T_{[e]}M$ by the standard averaging procedure and extend it to a G-invariant Q-Hermitian Riemannian metric h on M. More explicitly, we can construct such a scalar product $h_{[e]}$ on $T_{[e]}M \cong T_{[e]}M(V) \oplus W$ as orthogonal sum of K-invariant Euclidean scalar products on $T_{[e]}M(V)$ and W as follows. Using the open embedding $M(V) \hookrightarrow \tilde{M}(V)$ we can identify $T_{eK}M(V) \cong T_{e\tilde{K}}\tilde{M}(V) \cong \mathbb{R}^4 \otimes \mathbb{R}^{r,1} \cong \mathbb{R}^4 \otimes \mathbb{R}^{r+1}$ and choose the standard $\mathrm{O}(4) \times \mathrm{O}(r+1)$-invariant Euclidean scalar product on $\mathbb{R}^4 \otimes \mathbb{R}^{r+1}$. This scalar product is automatically K-invariant and $Q_{[e]}$-Hermitian. On W we choose any K-invariant Euclidean scalar product (which exists by compactness of K). It is automatically $Q_{[e]}$-Hermitian because $Q_{[e]} \subset \mathrm{End}(W)$ is precisely the image of $\mathfrak{o}(3) \subset \mathfrak{k} = \mathrm{Lie}\, K$ under the isotropy representation of $\mathfrak{k}$ on the $\mathfrak{k}$-invariant subspace $W \subset T_{[e]}M \cong T_{[e]}M(V) \oplus W$. It remains to show that $\mathrm{Isom}(M,h)$ does not contain any transitive solvable Lie subgroup if $r > 0$. In fact, M is homotopy equivalent to the simply connected real Grassmannian $\mathrm{SO}(3+r)/\mathrm{SO}(3) \times \mathrm{SO}(r)$ of oriented 3-planes in $\mathbb{R}^{3+r}$ $(r > 0)$. On the other hand, if (M,h) admits a transitive solvable group of isometries then M must be homotopy equivalent to a (possibly trivial) torus, which contradicts the fact that M is simply connected (and not contractible).
□

The last argument proves, in fact, the following theorem.

THEOREM 12. *Let $\mathfrak{p}(\Pi)$ be any extended Poincaré algebra of signature (p,q), $p > 3$ and $M = M(\Pi)$ the manifold constructed in Thm. 8. Then M does not admit any transitive (topological) action by a solvable Lie group.*

3. Bundles associated to the quaternionic manifold (M, Q)

To any almost quaternionic manifold (M, Q) one can canonically associate the following bundles over M: the twistor bundle $Z(M)$, the canonical SO(3)-principal bundle $S(M)$ and the Swann bundle $U(M)$. The **twistor bundle** (or **twistor space**) $Z(M) \to M$ is the subbundle of Q whose fibre $Z(M)_m$ at $m \in M$ consists of all complex structures subordinate to the quaternionic structure Q_m, i.e.

$$Z(M)_m = \{A \in Q_m | A^2 = -\mathrm{Id}\} .$$

So $Z(M)$ is a bundle of 2-spheres. The fibre $S(M)_m$ of the SO(3)-principal bundle $S(M)$ at $m \in M$ consists of all hypercomplex structures (J_1, J_2, J_3) subordinate to Q_m. Finally,

$$U(M) = S(M) \times_{\mathrm{SO}(3)} (\mathbb{H}^*/\{\pm 1\})$$

is associated to the action of $\mathrm{SO}(3) \cong \mathrm{Sp}(1)/\{\pm 1\}$ on $\mathbb{H}^*/\{\pm 1\}$ induced by left-multiplication of unit quaternions on $\mathbb{H}$. The total space $Z(M)$ carries a canonical almost complex structure $\mathbb{J}$, which is integrable if Q is quaternionic, see [**A-M-P**]. Similarly, one can define an almost hypercomplex structure $(\mathbb{J}_1, \mathbb{J}_2, \mathbb{J}_3)$ on $U(M)$, which is integrable if Q is quaternionic, cf. [**P-P-S**]. We recall the definition of the complex structure $\mathbb{J}$ on the twistor space $Z = Z(M)$ of a quaternionic manifold (M, Q). Since Q is 1-integrable, there exists a quaternionic connection ∇ on M, see Def. 6 and Def. 7. The holonomy of ∇ preserves not only $Q \subset \mathrm{End}(TM)$ but also its sphere subbundle $Z \subset Q$, simply because Id is a parallel section of $\mathrm{End}(TM)$. Let

$$TZ = TZ^{ver} \oplus TZ^{hor} \tag{17}$$

be the corresponding decomposition into the vertical space TZ^{ver} tangent to the fibres of the twistor bundle $Z \to M$ and its ∇-horizontal complement TZ^{hor}. The complex structure $\mathbb{J}$ preserves the decomposition (17). Let $m \in M$ be a point in M and $z := J \in Z_m \subset Q_m$ a complex structure on $T_m M$ subordinate to Q_m. Then $\mathbb{J}_z \in \mathrm{End}(T_z Z)$ is defined by:

$$\mathbb{J} A := JA \quad \text{and} \quad \mathbb{J}\tilde{X} = \widetilde{JX}$$

for all $A \in T_z Z^{ver} = T_z Z_m = \{A \in Q_m | AJ = -JA\}$ and all $X \in T_m M$, where $\tilde{X} \in T_z Z^{hor}$ denotes the ∇-horizontal lift of X. It was proven in [**A-M-P**] that $\mathbb{J}$ does not depend on the choice of quaternionic connection ∇.

If (M, Q) admits a quaternionic pseudo-Kähler metric g (of nonzero scalar curvature) then it is known that $(Z(M), \mathbb{J})$ admits a complex contact structure and a pseudo-Kähler-Einstein metric [**S1**], that $S(M)$ admits a pseudo-3-Sasakian structure [**Ko**] and that $(U(M), \mathbb{J}_1, \mathbb{J}_2, \mathbb{J}_3)$ admits a pseudo-hyper-Kähler metric [**Sw1**]. Moreover, all these special geometric structures are canonically associated to the data (M, Q, g). We recall that a **complex contact structure** on a complex manifold Z is a holomorphic distribution $\mathcal{D}$ of codimension one whose Frobenius form $[\cdot, \cdot] : \wedge^2 \mathcal{D} \to TZ/\mathcal{D}$ is (pointwise) nondegenerate; for the definition of 3-Sasakian structure see [**I-K**] and [**T**]. If a Lie group G acts (smoothly) on an almost quaternionic manifold (M, Q) preserving Q then there is an induced $\mathbb{J}$-holomorphic action on Z. Similarly, if a Lie group G acts on a quaternionic pseudo-Kähler manifold (M, Q, g) preserving the data (Q, g) then it acts on any of the bundles $Z(M)$, $S(M)$ and $U(M)$ preserving all the special geometric structures mentioned above.

THEOREM 13. *Let $(M(\Pi) = G(\Pi)/K, Q)$ be the homogeneous quaternionic manifold associated to an extended Poincaré algebra of signature (p,q), $p \geq 3$, $Z(\Pi) := Z(M(\Pi))$ its twistor space, $S(\Pi) := S(M(\Pi))$ its canonical $\mathrm{SO}(3)$-principal bundle and $U(\Pi) := U(M(\Pi))$ its Swann bundle. Then $G(\Pi)$ acts transitively on the manifolds $Z(\Pi)$ and $S(\Pi)$ and acts on $U(\Pi)$ with an orbit of codimension one.*

Proof: Since $G = G(\Pi)$ acts transitively on the base $M = M(\Pi)$ of any of the bundles $Z(\Pi) \to M$, $S(\Pi) \to M$ and $U(\Pi) \to M$, it is sufficient to consider the action of the stabilizer $K = \mathrm{Spin}(3) \cdot \mathrm{Spin}_0(r,q)$ on the fibres $Z(\Pi)_{[e]}$, $S(\Pi)_{[e]}$ and $U(\Pi)_{[e]}$, $[e] = eK \in M = G/K$. The subgroup $\mathrm{Spin}_0(r,q) \subset K$ acts trivially on $S(\Pi)_{[e]}$ and hence also on $Z(\Pi)_{[e]}$ and $U(\Pi)_{[e]}$, whereas $\mathrm{Spin}(3)$ acts transitively on the set $S(\Pi)_{[e]}$ of hypercomplex structures subordinate to $Q_{[e]}$ and hence also on the set $Z(\Pi)_{[e]}$ of complex structures subordinate to $Q_{[e]}$. From this it follows that G acts transitively on $Z(\Pi)$ and $S(\Pi)$ and with an orbit of codimension one on $U(\Pi)$. □

From now on we denote by $m_0 := [e] = eK \in M = G/K$ the canonical base point of M and fix the complex structure $J_1 \in Q_{m_0}$ as base point $z_0 := J_1 \in Z_{m_0}$ in $Z = Z(\Pi)$.

COROLLARY 7. *$(Z = Gz_0 \cong G/G_{z_0}, \mathbb{J})$ is a homogeneous complex manifold of the group G. The stabilizer of the point $z_0 \in Z$ in G is the centralizer of J_1 in K: $G_{z_0} = \mathrm{Z}_K(J_1) = \mathrm{Z}_{\mathrm{Spin}(3)}(J_1) \cdot \mathrm{Spin}_0(r,q)$, $\mathrm{Z}_{\mathrm{Spin}(3)}(J_1) \cong \mathrm{U}(1)$.*

Now we are going to construct a natural holomorphic immersion $Z \to \bar{Z} = \bar{Z}(\Pi) = G^{\mathbb{C}}/H$ of Z into a homogeneous complex manifold of the complexified linear group $G^{\mathbb{C}} \subset \mathrm{Aut}(\mathfrak{r}^{\mathbb{C}})$, where $(G_{z_0})^{\mathbb{C}} \subset H \subset G^{\mathbb{C}}$ are closed complex Lie subgroups.

First of all, we give an explicit description of the complex structure $\mathbb{J}$ on the twistor space Z. The choice of base point $z_0 \in Z$ determines a G-equivariant diffeomorphism $Z = Gz_0 \xrightarrow{\sim} G/G_{z_0}$, which maps z_0 to the canonical base point $eG_{z_0} \in G/G_{z_0}$. From now on we will identify Z and G/G_{z_0} via this map. The complex structure $\mathbb{J}$ being G-invariant, it is completely determined by the G_{z_0}-invariant complex structure $\mathbb{J}_{z_0}$ on $T_{z_0}Z$. In order to describe $\mathbb{J}_{z_0}$ we introduce the following G_{z_0}-invariant complement $\mathfrak{z} = \mathfrak{z}(\Pi)$ to $\mathfrak{g}_{z_0} = \mathrm{Lie}\, G_{z_0} = \mathbb{R}e_2 \wedge e_3 \oplus \mathfrak{o}(r,q)$ in $\mathfrak{g}$:

$$\mathfrak{z} = \mathbb{R}e_1 \wedge e_2 + \mathbb{R}e_1 \wedge e_3 + \mathfrak{m}\,.$$

The G_{z_0}-invariant decomposition $\mathfrak{g} = \mathfrak{g}_{z_0} + \mathfrak{z}$ determines a G_{z_0}-equivariant isomorphism $T_{z_0}Z \cong \mathfrak{g}/\mathfrak{g}_{z_0} \xrightarrow{\sim} \mathfrak{z}$. Using it we can consider the G_{z_0}-invariant complex structure $\mathbb{J}_{z_0}$ as a G_{z_0}-invariant complex structure on $\mathfrak{z}$.

PROPOSITION 11. *The G-invariant complex structure $\mathbb{J}$ on the twistor space $Z = G/G_{z_0}$ is given on $T_{z_0}Z \cong \mathfrak{z} = \mathbb{R}e_1 \wedge e_2 + \mathbb{R}e_1 \wedge e_3 + \mathfrak{m}$ by:*

$$\mathbb{J}_{z_0} e_1 \wedge e_2 = e_1 \wedge e_3\,, \quad \mathbb{J}_{z_0}|\mathfrak{m} = J_1\,.$$

Proof: Let ∇ be the G-invariant quaternionic connection on (M,Q) constructed in Lemma 6 and $L_x = \sum_{\alpha=1}^{3} \omega_\alpha(x) J_\alpha + \bar{L}_x$, $x \in \mathfrak{m}$, its Nomizu operators, where $\bar{L}_x \in \mathrm{z}(Q) \cong \mathfrak{gl}(d,\mathbb{H})$ $(d = \dim_{\mathbb{H}} M)$. The connection ∇ induces the decomposition $T_{z_0}Z = T_{z_0}Z^{ver} \oplus T_{z_0}Z^{hor} \cong \mathfrak{z} = \mathfrak{z}^{ver} \oplus \mathfrak{z}^{hor}$ into vertical space and horizontal space. The vertical space is $T_{z_0}Z^{ver} = T_{z_0}Z_{z_0} = \mathbb{R}J_2 \oplus \mathbb{R}J_3 \subset Q_{z_0}$ and $\mathfrak{z}^{ver} = \mathbb{R}e_1 \wedge e_2 \oplus \mathbb{R}e_1 \wedge e_3$ respectively, the identification being $J_2 \mapsto -e_1 \wedge e_2$, $J_3 \mapsto -e_1 \wedge e_3$.

For any vector $x \in \mathfrak{m}$ we consider the curve $c(t) = \exp txK \in G/K = M$ $(t \in \mathbb{R})$ and define a lift $s(t) \in Z_{c(t)}$ by the differential equation $\mathcal{L}_X s = 0$ with initial condition $s(0) = z_0 = J_1$. Here $X = \alpha(x)$ is the fundamental vector field on M associated to x (as defined on p. 22) and $\mathcal{L}_X$ is the Lie derivative with respect to X. Then $s(t) = (\exp tx)z_0$ is precisely the orbit of $z_0 \in Z = G/G_{z_0}$ under the 1-parameter subgroup of G generated by x. The vector

$$\begin{aligned} \frac{d}{dt}|_{t=0}(s - t\nabla_X s) &= s'(0) - \nabla_X s|_{t=0} \\ &= s'(0) + [L_x, J_1] = s'(0) - 2\omega_2(x)J_3 + 2\omega_3(x)J_2 \in T_{z_0}Z \end{aligned}$$

is horizontal. It is precisely the horizontal lift of $X(m_0) \in T_{m_0}M$ and corresponds to

$$\tilde{x} := x + 2\omega_2(x)e_1 \wedge e_3 - 2\omega_3(x)e_1 \wedge e_2 \in \mathfrak{z} \tag{18}$$

under the identification $T_{z_0}Z \cong \mathfrak{z}$. This shows that

$$\begin{aligned} \mathfrak{z}^{hor} &= \{\tilde{x} | x \in \mathfrak{m}\} \\ &= \mathbb{R}(e_2 + e_1 \wedge e_3) + \mathbb{R}(e_3 - e_1 \wedge e_2) + \\ &\quad \mathbb{R}e_0 + \mathbb{R}e_1 + \mathrm{span}\{e_{\iota a} | \iota = 0, \dots, 3,\ a = 1, \dots, r+q\} + W, \end{aligned}$$

see Lemma 6. Now the formulas for $\mathbb{J}_{z_0}$ follow easily. In fact, it is clear that $\mathbb{J}_{z_0}$ coincides with J_1 on the J_1-invariant subspace $\mathfrak{m} \cap \mathfrak{z}^{hor} = \mathbb{R}e_0 + \mathbb{R}e_1 + \mathrm{span}\{e_{\iota a} | \iota = 0, \dots, 3,\ a = 1, \dots, r+q\} + W = e_2^\perp \cap e_3^\perp \subset \mathfrak{m}$. The equation $\mathbb{J}_{z_0} e_1 \wedge e_2 = e_1 \wedge e_3$ follows immediately from $J_1 J_2 = J_3$, since $e_1 \wedge e_2,\ e_1 \wedge e_3 \in \mathfrak{z}^{ver}$ are identified with the vertical vectors $-J_2,\ -J_3 \in T_{z_0}Z^{ver}$. It is now sufficient to check that $\mathbb{J}_{z_0} e_2 = e_3$. This is done in the next computation:

$$\begin{aligned} \mathbb{J}_{z_0} e_2 &= \mathbb{J}_{z_0}(e_2 + e_1 \wedge e_3) - \mathbb{J}_{z_0}(e_1 \wedge e_3) = \mathbb{J}_{z_0}\tilde{e}_2 + e_1 \wedge e_2 \\ &= \widetilde{J_1 e_2} + e_1 \wedge e_2 = \tilde{e}_3 + e_1 \wedge e_2 = e_3 . \square \end{aligned}$$

We denote by $\mathfrak{z}^{1,0}$ (respectively, $\mathfrak{z}^{0,1}$) the eigenspace of $\mathbb{J}_{z_0} \in \mathrm{End}(\mathfrak{z})$ for the eigenvalue i (respectively, $-i$). The integrability of the complex structure $\mathbb{J}$ implies that $\mathfrak{h} := \mathfrak{h}(\Pi) := (\mathfrak{g}_{z_0})^{\mathbb{C}} + \mathfrak{z}^{0,1} \subset \mathfrak{g}^{\mathbb{C}}$ is a (complex) Lie subalgebra. Let $H = H(\Pi) \subset G^{\mathbb{C}} \subset \mathrm{Aut}(\mathfrak{r}^{\mathbb{C}})$ be the corresponding connected linear Lie group. Let us also consider the Lie algebra $\mathfrak{h}(V) := \mathfrak{h} \cap \mathfrak{g}(V)^{\mathbb{C}} = (\mathfrak{g}_{z_0})^{\mathbb{C}} + \mathfrak{z}(V)^{0,1} \subset \mathfrak{g}(V)^{\mathbb{C}}$, where $\mathfrak{z}(V) = \mathfrak{z} \cap \mathfrak{g}(V)$ and $\mathfrak{z}(V)^{0,1} := \mathfrak{z}^{0,1} \cap \mathfrak{g}(V)^{\mathbb{C}}$.

PROPOSITION 12. $H = H(\Pi) \subset G^{\mathbb{C}} \subset \mathrm{Aut}(\mathfrak{r}^{\mathbb{C}})$ *are complex algebraic subgroups.*

Proof: It follows from Prop. 4 that $\mathfrak{g}^{\mathbb{C}} \subset \mathrm{der}(\mathfrak{r}^{\mathbb{C}})$ is a complex algebraic subalgebra and hence $G^{\mathbb{C}} \subset \mathrm{Aut}(\mathfrak{r}^{\mathbb{C}})$ a complex algebraic subgroup. It only remains to show that $\mathfrak{h}$ is a complex algebraic subalgebra. Let us consider the decomposition $\mathfrak{h} = (\mathfrak{g}_{z_0})^{\mathbb{C}} + \mathfrak{z}^{0,1}$. The subalgebra $(\mathfrak{g}_{z_0})^{\mathbb{C}}$ is algebraic. In fact, $\mathfrak{g}_{z_0} = z_{\mathfrak{k}}(e_2 \wedge e_3)$ is a centralizer in the real algebraic subalgebra $\mathfrak{k} \subset \mathfrak{g}$. If the subalgebra $\langle \mathfrak{z}^{0,1} \rangle \subset \mathfrak{h}$ generated by the subspace $\mathfrak{z}^{0,1}$ is an algebraic subalgebra of $\mathrm{der}(\mathfrak{r}^{\mathbb{C}})$, then $\mathfrak{h}$ is generated by algebraic linear Lie algebras and hence is itself algebraic, see **[O-V]**. The algebraicity of $\langle \mathfrak{z}^{0,1} \rangle$ is proven in the next lemma. $\square$

LEMMA 7. $\mathfrak{z}^{0,1}$ *generates the algebraic subalgebra* $\langle \mathfrak{z}^{0,1} \rangle = \mathfrak{o}(E')^{\mathbb{C}} + \mathfrak{z}^{0,1} \subset \mathfrak{g}^{\mathbb{C}}$.

Proof: First we compute the subalgebra $\langle \mathfrak{z}^{0,1} \rangle$ of $\mathfrak{g}^{\mathbb{C}}$ generated by $\mathfrak{z}^{0,1} = \mathfrak{z}^{0,1}(V) + W^{0,1}$. Note that

$$\begin{aligned} \mathfrak{z}^{0,1}(V) &= \mathrm{span}\{e_1 \wedge e_2 + ie_1 \wedge e_3, e_0 + ie_1, e_2 + ie_3\} \\ &\quad + \mathrm{span}\{e_{0a} + ie_{1a}, e_{2a} + ie_{3a} | a = 1, \ldots, r+q\}. \end{aligned}$$

It is easy to check that

$$\begin{aligned} [\mathfrak{z}^{0,1}(V), \mathfrak{z}^{0,1}(V)] &= \mathfrak{o}(E')^{\mathbb{C}} + \mathrm{span}\{e_1 \wedge e_2 + ie_1 \wedge e_3, e_2 + ie_3\} \\ &\quad + \mathrm{span}\{e_{2a} + ie_{3a} | a = 1, \ldots, r+q\} \end{aligned}$$

and

$$[\mathfrak{z}^{0,1}(V), W^{0,1}] = W^{0,1}.$$

We show that $[W^{0,1}, W^{0,1}] \subset \mathbb{C}(e_2 + ie_3)$. This shows that $\langle \mathfrak{z}^{0,1} \rangle \supset \mathfrak{o}(E')^{\mathbb{C}} + \mathfrak{z}^{0,1}$. Since the right-hand side is closed under Lie brackets, we conclude that $\langle \mathfrak{z}^{0,1} \rangle = \mathfrak{o}(E')^{\mathbb{C}} + \mathfrak{z}^{0,1}$. It is easy to check that the subalgebra

$$\begin{aligned} \mathfrak{o}(E')^{\mathbb{C}} &+ \mathrm{span}\{e_1 \wedge e_2 + ie_1 \wedge e_3, e_2 + ie_3\} \\ &+ \mathrm{span}\{e_{0a} + ie_{1a}, e_{2a} + ie_{3a} | a = 1, \ldots, r+q\} + W^{0,1} \end{aligned}$$

coincides with its derived Lie algebra. Therefore, it is an algebraic subalgebra of $\mathfrak{g}^{\mathbb{C}}$, see [**O-V**]. Now, to prove that $\langle \mathfrak{z}^{0,1} \rangle$ is algebraic, it is sufficient to check that $\mathbb{C}(e_0 + ie_1)$ is algebraic. The element $e_0 + ie_1 \in \mathfrak{g}^{\mathbb{C}}$ is conjugated to the algebraic element $e_0 \in \mathfrak{g}^{\mathbb{C}}$ via $\exp(-ie_1) \in G^{\mathbb{C}}$. Consequently, it is algebraic. □

THEOREM 14. *For the twistor space Z of any of the homogeneous quaternionic manifolds $(M = G/K, Q)$ constructed in Thm. 8 there is a natural open G-equivariant holomorphic immersion*

$$Z \cong G/G_{z_0} \to \bar{Z} = \bar{Z}(\Pi) := G^{\mathbb{C}}/H$$

into a homogeneous complex (Hausdorff) manifold of the complex algebraic group $G^{\mathbb{C}}$. This immersion is a (universal) finite covering over its image, which is an open G-orbit. (The group H was defined on p. 36.)

Proof: It follows from Prop. 12 that $H \subset G^{\mathbb{C}}$ is closed. This shows that $G^{\mathbb{C}}/H$ is a homogeneous complex Hausdorff manifold. The inclusions $G \subset G^{\mathbb{C}}$ and $G_{z_0} \subset H$ define a G-equivariant map $G/G_{z_0} \to G^{\mathbb{C}}/H$, which is an immersion since $\mathfrak{g} \cap \mathfrak{h} = \mathfrak{g}_{z_0}$. The differential of this immersion at eG_{z_0} is canonically identified with the restriction $\phi : \mathfrak{z} \to \mathfrak{g}^{\mathbb{C}}/\mathfrak{h}$ of the canonical projection $\mathfrak{g}^{\mathbb{C}} \to \mathfrak{g}^{\mathbb{C}}/\mathfrak{h}$ to $\mathfrak{z} \subset \mathfrak{g} = \mathfrak{g}_{z_0} + \mathfrak{z}$. Obviously, the complex linear extension $\phi^{\mathbb{C}}$ maps $\mathfrak{z}^{1,0}$ isomorphically to $\mathfrak{g}^{\mathbb{C}}/\mathfrak{h}$ and $\mathfrak{z}^{0,1}$ to zero. This shows that $Z \cong G/G_{z_0} \to G^{\mathbb{C}}/H$ is open and holomorphic with respect to the G-invariant complex structure $\mathbb{J}$ on Z and the canonical complex structure on $G^{\mathbb{C}}/H$. □

THEOREM 15. *The homogeneous complex manifold $\bar{Z} = G^{\mathbb{C}}/H$ carries a $G^{\mathbb{C}}$-invariant holomorphic hyperplane distribution $\mathcal{D} \subset T\bar{Z}$. The hyperplane $\mathcal{D}_{z_0} = T_{z_0}Z^{hor} \subset T_{z_0}Z = T_{z_0}\bar{Z}$ is the horizontal space associated to the G-invariant quaternionic connection ∇ on M constructed in Lemma 6. Moreover, $\mathcal{D}$ defines a complex contact structure on $\bar{Z} = \bar{Z}(\Pi)$ if and only if Π is nondegenerate. In this* *case the restriction $\mathcal{D}|Z$ coincides with the canonical complex contact structure on the twistor space Z of the quaternionic pseudo-Kähler manifold M.*

Proof: Recall that we identify $T_{z_0}Z$ with the $\mathfrak{g}_{z_0}$-invariant subspace $\mathfrak{z} \subset \mathfrak{g} = \mathfrak{g}_{z_0} + \mathfrak{z}$ complementary to $\mathfrak{g}_{z_0}$. Hereby the subspace $\mathcal{D}_{z_0} = T_{z_0}Z^{hor}$ is identified with $\mathfrak{z}^{hor} = \{\tilde{x}|x \in \mathfrak{m}\} \subset \mathfrak{z}$, where $\tilde{x} = x + 2\omega_2(x)e_1 \wedge e_3 - 2\omega_3(x)e_1 \wedge e_2$ is the ∇-horizontal lift of x, see equation (18). The subspace $\mathfrak{z}^{hor} \subset \mathfrak{z}$ is $\mathbb{J}_{z_0}$-invariant by the very definition of the complex structure $\mathbb{J}_{z_0}$ and $\mathbb{J}_{z_0}|\mathfrak{z}^{hor}$ is given by $\mathbb{J}_{z_0}\tilde{x} = \widetilde{J_1 x}$ for all $x \in \mathfrak{m}$. The subspace $(\mathfrak{z}^{hor})^{1,0} = (\mathfrak{z}^{hor})^{\mathbb{C}} \cap \mathfrak{z}^{1,0} \subset \mathfrak{z}^{1,0}$ is identified with the i-eigenspace $\mathcal{D}^{1,0}_{z_0} \subset T^{1,0}_{z_0}Z = T^{1,0}_{z_0}\bar{Z}$ of $\mathbb{J}_{z_0}$ on $\mathcal{D}_{z_0}$. In order to prove that $\mathcal{D}^{1,0}_{z_0}$ extends to a $G^{\mathbb{C}}$-invariant holomorphic distribution $\mathcal{D}^{1,0} \subset T^{1,0}\bar{Z}$ it is sufficient to check the following lemma.

LEMMA 8. *The complex Lie algebra $\mathfrak{h} = \mathfrak{g}^{\mathbb{C}}_{z_0} + \mathfrak{z}^{0,1}$ preserves the projection of $\mathfrak{z}^{hor})^{1,0} = \{\tilde{x} - i\widetilde{J_1 x}|x \in \mathfrak{m}\} = \mathrm{span}_{\mathbb{C}}\{e_2 + e_1 \wedge e_3 - i(e_3 - e_1 \wedge e_2), e_0 - ie_1, e_{0a} - ie_{1a}, e_{2a} - ie_{3a}|a = 1, \ldots, r+q\} + W^{1,0}$ into $\mathfrak{g}^{\mathbb{C}}/\mathfrak{h}$, i.e.*

$$[\mathfrak{h}, (\mathfrak{z}^{hor})^{1,0}] \subset \mathfrak{h} + (\mathfrak{z}^{hor})^{1,0}\,.$$

Now $\mathcal{D}^{1,0}$ defines a complex contact structure if and only if the Frobenius form $\wedge^2\mathcal{D}^{1,0} \overset{[\cdot,\cdot]}{\to} T^{1,0}\bar{Z}/\mathcal{D}^{1,0}$ is nondegenerate, which is equivalent to the nondegeneracy of the skew symmetric complex bilinear form

$$\omega : \wedge^2(\mathfrak{z}^{hor})^{1,0} \overset{[\cdot,\cdot]}{\to} \mathfrak{g}^{\mathbb{C}} \to \mathfrak{g}^{\mathbb{C}}/(\mathfrak{h} + (\mathfrak{z}^{hor})^{1,0}) \cong \mathbb{C}\,.$$

LEMMA 9. *Let $(\mathfrak{z}^{hor})^{1,0} = (\mathfrak{z}(V)^{hor})^{1,0} + W^{1,0}$ be the decomposition of $(\mathfrak{z}^{hor})^{1,0}$ induced by the decomposition $\mathfrak{z} = \mathfrak{z}(V) + W$. Then $\omega((\mathfrak{z}(V)^{hor})^{1,0}, W^{1,0}) = 0$, $\omega| \wedge^2 (\mathfrak{z}(V)^{hor})^{1,0}$ is nondegenerate and $\omega| \wedge^2 W^{1,0}$ is given by*

$$\omega(s - iJ_1 s, t - iJ_1 t) = 2(\langle e_2, [s,t]\rangle + i\langle e_3, [s,t]\rangle)(e_2 - ie_3) \pmod{\mathfrak{h} + (\mathfrak{z}^{hor})^{1,0}}\,. \tag{19}$$

From the lemma it follows that ω is nondegenerate if and only if $\omega| \wedge^2 W^{1,0}$ is nondegenerate. The explicit formula for $\omega|\wedge^2 W^{1,0}$ given in equation (19) now shows that ω is nondegenerate if and only if $b(\Pi) = b_{\Pi,(e_1,e_2,e_3)}$ is nondegenerate, which is in turn equivalent to the nondegeneracy of Π by Cor. 1. This proves that $\mathcal{D}$ is a complex contact structure if and only if Π is nondegenerate. If Π is nondegenerate then $\mathcal{D}_{z_0} = T_{z_0}Z^{hor}$ is precisely the horizontal space associated to the Levi-Civita connection ∇^g of the quaternionic pseudo-Kähler manifold (M, Q, g) and $\mathcal{D}$ is the canonical complex contact structure on its twistor space Z, as defined by Salamon **[S1]**. □

4. Homogeneous quaternionic supermanifolds associated to superextended Poincaré algebras

In this section we will show that our main result, Thm. 8, has a natural supergeometric analogue, Thm. 17. The fundamental idea is to replace the map $\Pi : \wedge^2 W \to V$ defined on the exterior square $\wedge^2 W$ by an $\mathfrak{o}(V)$-equivariant linear map $\Pi : \vee^2 W \to V$ defined on the symmetric square $\vee^2 W = \mathrm{Sym}^2 W$. We will freely use the language of supergeometry. The necessary background is outlined in the appendix.

4.1. Superextended Poincaré algebras. Let $(V, \langle \cdot, \cdot \rangle)$ be a pseudo-Euclidean vector space, W a $C\ell^0(V)$-module and $\Pi : \vee^2 W \to V$ an $\mathfrak{o}(V)$-equivariant linear map. We recall that $\mathfrak{o}(V)$ acts on W via $\mathrm{ad}^{-1} : \mathfrak{o}(V) \to \mathfrak{spin}(V) \subset C\ell^0(V)$, see equation (2).

Given these data we extend the Lie bracket on $\mathfrak{p}_0 := \mathfrak{p}(V)$ to a super Lie bracket (see Def. 16) $[\cdot, \cdot]$ on the $\mathbb{Z}_2$-graded vector space $\mathfrak{p}_0 + \mathfrak{p}_1$, $\mathfrak{p}_1 = W$, by the following requirements:

1) The adjoint representation (see Def. 28) of $\mathfrak{o}(V)$ on $\mathfrak{p}_1$ concides with the natural representation of $\mathfrak{o}(V) \cong \mathfrak{spin}(V)$ on $W = \mathfrak{p}_1$ and $[V, \mathfrak{p}_1] = 0$.
2) $[s, t] = \Pi(s \vee t)$ for all $s, t \in W$.

The super Jacobi identity follows from 1) and 2). The resulting super Lie algebra will be denoted by $\mathfrak{p}(\Pi)$.

Definition 13. *Any super Lie algebra $\mathfrak{p}(\Pi)$ as above is called a* **superextended Poincaré algebra** *(of* **signature** (p,q) *if $V \cong \mathbb{R}^{p,q}$). $\mathfrak{p}(\Pi)$ is called* **nondegenerate** *if Π is nondegenerate, i.e. if the map $W \ni s \mapsto \Pi(s \vee \cdot) \in W^* \otimes V$ is injective.*

The structure of superextended Poincaré algebra on the vector space $\mathfrak{p}(V) + W$ is completely determined by the map $\Pi : \vee^2 W \to V$. An explicit basis for the vector space $(\vee^2 W^* \otimes V)^{\mathfrak{o}(V)}$ of such $\mathfrak{o}(V)$-equivariant linear maps was constructed in [**A-C2**] for all V and W.

4.2. The canonical supersymmetric bilinear form b. Let $V = \mathbb{R}^{p,q}$ be the standard pseudo-Euclidean vector space with scalar product $\langle \cdot, \cdot \rangle$ of signature (p, q). From now on we fix a decomposition $p = p' + p''$ and assume that $p' \equiv 3 \pmod 4$, see Remark 7 below. We denote by $(e_i) = (e_1, \ldots, e_{p'})$ the first p' basis vectors of the standard basis of V and by $(e'_i) = (e'_1, \ldots, e'_{p''+q})$ the remaining ones. The two complementary orthogonal subspaces of V spanned by these bases are denoted by $E = \mathbb{R}^{p'} = \mathbb{R}^{p',0}$ and $E' = \mathbb{R}^{p'',q}$ respectively. The vector spaces V, E and E' are oriented by their standard orthonormal bases. E.g. the orientation of Euclidean p'-space E defined by the basis (e_i) is $e_1^* \wedge \cdots \wedge e_{p'}^* \in \wedge^{p'} E^*$. Here (e_i^*) denotes the basis of E^* dual to (e_i). Now let $\mathfrak{p}(\Pi) = \mathfrak{p}(V) + W$ be a superextended Poincaré algebra of signature (p, q) and $(\tilde{e}_i)$ any orhonormal basis of E. Then we define a $\mathbb{R}$-bilinear form $b_{\Pi,(\tilde{e}_i)}$ on the $C\ell^0(V)$-module W by:

$$b_{\Pi,(\tilde{e}_i)}(s,t) = \langle \tilde{e}_1, [\tilde{e}_2 \ldots \tilde{e}_{p'} s, t] \rangle = \langle \tilde{e}_1, \Pi(\tilde{e}_2 \ldots \tilde{e}_{p'} s \vee t) \rangle, \quad s, t \in W. \tag{20}$$

We put $b = b(\Pi) := b_{\Pi,(e_i)}$ for the standard basis (e_i) of E. As in 4.1 we consider $W = \mathfrak{p}_1$ as $\mathbb{Z}_2$-graded vector space of purely odd degree and recall that, by Def. 22, an even supersymmetric (respectively, super skew symmetric) bilinear form on W

is simply an ordinary skew symmetric (respectively, symmetric) bilinear form on W.

Remark 7: Equation (20) defines an even super skew symmetric bilinear form on W if $p' \equiv 1 \pmod 4$. For even p' the above formula does not make sense, unless one assumes that W is a $C\ell(V)$-module rather than a $C\ell^0(V)$-module. Here we are only interested in the case $p' \equiv 3 \pmod 4$. Moreover, later on, for the construction of homogeneous quaternionic supermanifolds we will put $p' = 3$.

THEOREM 16. *The bilinear form b has the following properties:*

1) $b_{\Pi,(\tilde{e}_i)} = \pm b$ *if* $\tilde{e}_1 \wedge \cdots \wedge \tilde{e}_{p'} = \pm e_1 \wedge \cdots \wedge e_{p'}$. *In particular, our definition of b does not depend on the choice of positively oriented orthonormal basis of E.*
2) *b is an even supersymmetric bilinear form.*
3) *b is invariant under the connected subgroup* $K(p', p'') = \mathrm{Spin}(p')\cdot\mathrm{Spin}_0(p'', q) \subset \mathrm{Spin}_0(p, q)$ *(and is not* $\mathrm{Spin}_0(p, q)$*-invariant, unless* $p'' + q = 0$*).*
4) *Under the identification* $\mathfrak{o}(V) = \wedge^2 V = \wedge^2 E + \wedge^2 E' + E \wedge E'$, *see equation (3), the subspace $E \wedge E'$ acts on W by b-symmetric endomorphisms and the subalgebra* $\wedge^2 E \oplus \wedge^2 E' \cong \mathfrak{o}(p') \oplus \mathfrak{o}(p'', q)$ *acts on W by b-skew symmetric endomorphisms.*

Proof: The proof is the same as for Thm. 1, up to the modifications caused by the fact that Π is now symmetric instead of skew symmetric. Part 2) e.g. follows from the next computation, in which we use that $p' \equiv 3 \pmod 4$:

$$\begin{aligned} b(t,s) &= \langle e_1, [e_2 \ldots e_{p'} t, s] \rangle \\ &= -\langle e_1, [e_4 \ldots e_{p'} t, e_2 e_3 s] \rangle + \langle e_1, \mathrm{ad}(e_2 e_3)[e_4 \ldots e_{p'} t, s] \rangle \\ &= -\langle e_1, [e_4 \ldots e_{p'} t, e_2 e_3 s] \rangle = \cdots = -\langle e_1, [t, e_2 \ldots e_{p'} s] \rangle \\ &= -\langle e_1, [e_2 \ldots e_{p'} s, t] \rangle = -b(s,t) \,.\square \end{aligned}$$

DEFINITION 14. *The bilinear form* $b = b(\Pi) = b_{\Pi,(e_1,\ldots,e_{p'})}$ *defined above is called the* **canonical supersymmetric bilinear form** *on W associated to the* $\mathfrak{o}(V)$*-equivariant map* $\Pi : \vee^2 W \to V = \mathbb{R}^{p,q}$ *and the decomposition* $p = p' + p''$.

PROPOSITION 13. *The kernels of the linear maps* $\Pi : W \to W^* \otimes V$ *and* $b = b(\Pi) : W \to W^*$ *coincide:* $\ker \Pi = \ker b$.

Proof: See the proof of Prop. 1. □

COROLLARY 8. $\mathfrak{p}(\Pi)$ *is nondegenerate (see Def. 13) if and only if* $b(\Pi)$ *is nondegenerate.*

4.3. The main theorem in the super case. Any superextended Poincaré algebra $\mathfrak{p} = \mathfrak{p}(\Pi) = \mathfrak{p}(V) + W$ has an even derivation D with eigenspace decomposition $\mathfrak{p} = \mathfrak{o}(V) + V + W$ and corresponding eigenvalues $(0, 1, 1/2)$. Therefore, the super Lie algebra $\mathfrak{p} = \mathfrak{p}_0 + \mathfrak{p}_1 = \mathfrak{p}(V) + W$ is canonically extended to a super Lie algebra $\mathfrak{g} = \mathfrak{g}(\Pi) = \mathbb{R}D + \mathfrak{p} = \mathfrak{g}_0 + \mathfrak{g}_1$, where $\mathfrak{g}_0 = \mathbb{R}D + \mathfrak{p}_0 = \mathbb{R}D + \mathfrak{p}(V) = \mathfrak{g}(V)$ and $\mathfrak{g}_1 = \mathfrak{p}_1 = W$.

PROPOSITION 14. *The adjoint representation (see Def. 28) of $\mathfrak{g}$ is faithful and moreover it induces a faithful representation on its ideal* $\mathfrak{r} = \mathbb{R}D + V + W \subset \mathfrak{g} = \mathfrak{o}(V) + \mathfrak{r}$.

By Prop. 14 we can consider $\mathfrak{g}$ as subalgebra of the super Lie algebra $\mathfrak{gl}(\mathfrak{r})$ (defined in 4.6), i.e. $\mathfrak{g}$ is a linear super Lie algebra (see Def. 28). We denote by $G = G(\Pi)$ the corresponding linear Lie supergroup, see 4.6. Its underlying Lie group is $G_0 = G(V)$ and has $\mathfrak{g}_0 = \mathfrak{g}(V)$ as Lie algebra. For the construction of homogeneous quaternionic supermanifolds we will assume that $V = \mathbb{R}^{p,q}$ with $p \geq 3$. Then we fix the decomposition $p = p' + p''$, where now $p' = 3$ and $p'' = p - 3 =: r$. As before, we have a corresponding orthogonal decomposition $V = E + E'$, the subalgebra $\mathfrak{k} = \mathfrak{k}(p', p'') = \mathfrak{k}(3, r) = \mathfrak{o}(3) \oplus \mathfrak{o}(r, q) \subset \mathfrak{o}(p, q) = \mathfrak{o}(V)$ preserving this decomposition and the corresponding linear Lie subgroup $K = K(3, r) \subset G_0$. We are interested in the homogeneous supermanifold (see 4.7):

$$M = M(\Pi) := G/K = G(\Pi)/K\,.$$

Its underlying manifold is the homogeneous manifold

$$M_0 := G_0/K = G(V)/K = M(V)\,.$$

THEOREM 17. 1) *There exists a G-invariant quaternionic structure Q on $M = G/K$.*

2) *If Π is nondegenerate (see Def. 13) then there exists a G-invariant pseudo-Riemannian metric g on M such that (M, Q, g) is a quaternionic pseudo-Kähler supermanifold.*

Proof: First let (Q_0, g_0) denote the G_0-invariant quaternionic pseudo-Kähler structure on $M_0 = G_0/K$ which was introduced in the proof of Thm. 8 (previously it was denoted simply by (Q, g)). As in that proof, see equation (13), the corresponding K-invariant quaternionic structure on $T_{eK}M_0 \cong \mathfrak{g}_0/\mathfrak{k}$ is extended to a K-invariant quaternionic structure Q_{eK} on $T_{eK}M \cong \mathfrak{g}/\mathfrak{k}$ defining a G-invariant almost quaternionic structure Q on G/K (see 4.5 and 4.7). Similarly, the G_0-invariant pseudo-Riemannian metric g_0 on M_0 corresponds to a K-invariant pseudo-Euclidean scalar product on $\mathfrak{g}_0/\mathfrak{k}$. This scalar product is extended by $b = b_{\Pi,(e_1,e_2,e_3)}$, in the obvious way, to a K-invariant supersymmetric bilinear form g_{eK} on $\mathfrak{g}/\mathfrak{k}$, which is nondegenerate if Π is nondegenerate. Let us first treat the case where Π is nondegenerate. In this case g_{eK} defines a G-invariant Q-Hermitian pseudo-Riemannian metric g on G/K. The Levi-Civita connection of the homogeneous pseudo-Riemannian supermanifold (M, g) is computed in Lemma 10 below. As on p. 26, $\mathfrak{g}/\mathfrak{k}$ is identified with the complement $\mathfrak{m} = \mathfrak{m}(V) + W$ to $\mathfrak{k}$ in $\mathfrak{g}$. We use the g_0-orthonormal basis $(e_{\iota i})$ of $\mathfrak{m}(V)$ introduced on p. 27 and recall that $\epsilon_i = g_0(e_{\iota i}, e_{\iota i})$. Also we will continue to write $x \wedge_g y$ for the g_{eK}-skew symmetric endomorphism of $\mathfrak{m}$ defined for $x, y \in \mathfrak{m}(V)$ by: $x \wedge_g y(z) := g_{eK}(y, z)x - g_{eK}(x, z)y$, $z \in \mathfrak{m}$.

LEMMA 10. *The Nomizu map $L^g = L(\nabla^g)$ associated to the Levi-Civita connection ∇^g of the homogeneous pseudo-Riemannian supermanifold $(M = G/K, g)$ is given by the following formulas:*

$$\begin{aligned} L^g_{e_0} &= L^g_{e_{\alpha a}} = 0\,, \\ L^g_{e_\alpha} &= \frac{1}{2}J_\alpha + \bar{L}^g_{e_\alpha}\,, \end{aligned}$$

where

$$\bar{L}^g_{e_\alpha} = \frac{1}{2}\sum_{i=0}^{r+q} \epsilon_i e_{\alpha i} \wedge_g e'_i - \frac{1}{2}\sum_{i=0}^{r+q} \epsilon_i e_{\gamma i} \wedge_g e_{\beta i} \in \mathrm{z}(Q_{eK})\,,$$

$L^g_{e'_a} \in \mathrm{z}(Q_{eK})$ *is given by:*

$$\begin{aligned} L^g_{e'_a}|\mathfrak{m}(V) &= \sum_{\iota=0}^{3} e_{\iota a} \wedge_g e_\iota \,, \\ L^g_{e'_a}|W &= \frac{1}{2} e_1 e_2 e_3 e'_a \,. \end{aligned}$$

For all $s \in W$ the Nomizu operator $L^g_s \in \mathrm{z}(Q_{eK})$ maps the subspace $\mathfrak{m}(V) \subset \mathfrak{m} = \mathfrak{m}(V) + W$ into W and W into $\mathfrak{m}(V)$. The restriction $L^g_s|W$ $(s \in W)$ is completely determined by $L^g_s|\mathfrak{m}(V)$ (and vice versa) according to the relation

$$g_{eK}(L^g_s t, x) = g_{eK}(t, L^g_s x)\,, \quad s, t \in W\,, \quad x \in \mathfrak{m}(V)\,.$$

Finally, $L^g_s|\mathfrak{m}(V)$ $(s \in W)$ is completely determined by its values on the quaternionic basis (e'_i), $i = 0, \ldots, r+q$, which are as follows:

$$\begin{aligned} L^g_s e'_0 &= \frac{1}{2} s \,, \\ L^g_s e'_a &= \frac{1}{2} e_1 e_2 e_3 e'_a s \,. \end{aligned}$$

(It is understood that $L^g_x = \mathrm{ad}_x|\mathfrak{m}$ for all $x \in \mathfrak{k}$, cf. equation (24).) In the above formulas $a = 1, \ldots, r+q$, $\alpha = 1, 2, 3$ and (α, β, γ) is a cyclic permutation of $(1, 2, 3)$.

Proof: This follows from equation (26) by a straightforward computation. □

COROLLARY 9. *The Levi-Civita connection ∇^g of the homogeneous almost quaternionic pseudo-Hermitian supermanifold (M, Q, g) preserves Q and hence (M, Q, g) is a quaternionic pseudo-Kähler supermanifold.*

By Cor. 9 we have already established part 2) of Thm. 17. Part 1) is a consequence of 2) provided that Π is nondegenerate. It remains to discuss the case of degenerate Π. From the $\mathfrak{o}(V)$-equivariance of Π it follows that $W_0 = \ker \Pi \subset W$ is an $\mathfrak{o}(V)$-submodule. Let W' be a complementary $\mathfrak{o}(V)$-submodule. Then W_0 and W' are $C\ell^0(V)$-submodules; we put $\Pi' := \Pi| \vee^2 W'$. We denote by $(M' := M(\Pi'), Q', g')$ the corresponding quaternionic pseudo-Kähler supermanifold and by $L' := L^{g'} : \mathfrak{g}(\Pi') \to \mathrm{End}(\mathfrak{m}(\Pi'))$ the Nomizu map associated to its Levi-Civita connection. By the next lemma, we can extend the map L' to a torsionfree Nomizu map $L : \mathfrak{g}(\Pi) \to \mathrm{End}(\mathfrak{m}(\Pi))$, whose image normalizes Q_{eK}. This proves the 1-integrability of Q (by Cor. 10), completing the proof of Thm. 17. □

By Cor. 9 we can decompose $L'_x = \sum_{\alpha=1}^{3} \omega'_\alpha(x) J_\alpha + \bar{L}'_x$, where the ω'_α are 1-forms on $\mathfrak{m}(\Pi')$ and $\bar{L}'_x \in \mathrm{z}({Q'}_{eK})$ belongs to the centralizer of the quaternionic structure ${Q'}_{eK} = Q_{eK}|\mathfrak{m}(\Pi')$ on $\mathfrak{m}(\Pi')$.

LEMMA 11. *The Nomizu map $L' : \mathfrak{g}(\Pi') \to \mathrm{End}(\mathfrak{m}(\Pi'))$ associated to the Levi-Civita connection of $(M(\Pi') = G(\Pi')/K, g')$ can be extended to the Nomizu map $L : \mathfrak{g}(\Pi) \to \mathrm{End}(\mathfrak{m}(\Pi))$ of a $G(\Pi)$-invariant quaternionic connection ∇ on the homogeneous almost quaternionic supermanifold $(M(\Pi) = G(\Pi)/K, Q)$. The extension is defined as follows:*

$$L_x := \sum_{\alpha=1}^{3} \omega_\alpha(x) J_\alpha + \bar{L}_x \,, \quad x \in \mathfrak{m}(\Pi)\,,$$

where $\bar{L}_x \in \mathrm{z}(Q)$ *(the centralizer is taken in* $\mathfrak{gl}(\mathfrak{m}(\Pi))$*) is defined below and the 1-forms* ω_α *on* $\mathfrak{m}(\Pi) := \mathfrak{m}(\Pi') + W_0$ *satisfy* $\omega_\alpha|\mathfrak{m}(\Pi') := \omega'_\alpha$ *and* $\omega_\alpha|W_0 := 0$. *The operators* $\bar{L}_x$ *are given by:*

$$\bar{L}_x|\mathfrak{m}(\Pi') := \bar{L}'_x \quad \textit{if} \quad x \in \mathfrak{m}(\Pi')\,,$$

$$\bar{L}_{e_0}|W_0 := \bar{L}_{e_{\alpha a}}|W_0 := \bar{L}_{e_\alpha}|W_0 := 0\,,$$

$$\bar{L}_{e'_a}|W_0 := \frac{1}{2}e_1e_2e_3e'_a\,,$$

$$\bar{L}_s|W_0 := 0 \quad \textit{if} \quad s \in W\,,$$

$$\bar{L}_s x := (-1)^{\tilde{x}}L_x s - [s,x] \quad \textit{if} \quad s \in W_0\,, \quad x \in \mathfrak{m}(\Pi')\,.$$

(It is understood that $L_x = \mathrm{ad}_x|\mathfrak{m}(\Pi)$ *for all* $x \in \mathfrak{k}$.*) In the above formulas, as usual,* $a = 1, \ldots, r+q$, $\alpha = 1,2,3$ *and* $\tilde{x} \in \mathbb{Z}_2 = \{0,1\}$ *stands for the* $\mathbb{Z}_2$*-degree of* $x \in \mathfrak{m}(\Pi') = \mathfrak{m}(\Pi')_0 + \mathfrak{m}(\Pi')_1 = \mathfrak{m}(V) + W'$.

Proof: The proof is similar to that of Lemma 6. □

Appendix. Supergeometry

In this appendix we summarize the supergeometric material needed in 4. Standard references on supergeometry are **[M]**, **[L]** and **[K]**, see also **[B-B-H]**, **[Ba]**, **[Be]**, **[Bern]**, **[B-O]**, **[DW]**, **[F]**, **[O1]**, **[O2]**, **[Sch]** and **[S-W]**. (D.A. Leites has informed us that he will soon publish a monograph on supergeometry.)

4.4. Supermanifolds. Let $V = V_0 + V_1$ be a $\mathbb{Z}_2$-graded vector space. We recall that an element $x \in V$ is called **homogeneous** (or **of pure degree**) if $x \in V_0 \cup V_1$. The **degree** of a homogeneous element $x \in V$ is the number $\tilde{x} \in \mathbb{Z}_2 = \{0, 1\}$ such that $x \in V_{\tilde{x}}$. The element $x \in V$ said to be even if $\tilde{x} = 0$ and odd if $\tilde{x} = 1$. For V_0 and V_1 of finite dimension, the **dimension** of V is defined as $\dim V := \dim V_0 | \dim V_1 = m|n$ and a **basis** of V is by definition a tuple $(x_1, \ldots, x_m, \xi_1, \ldots, \xi_n)$ such that $(x_1, \ldots, x_m)$ is a basis of V_0 and $(\xi_1, \ldots, \xi_n)$ is a basis of V_1.

DEFINITION 15. *Let* $\mathbf{A}$ *be a* $\mathbb{Z}_2$*-graded algebra. The* **supercommutator** *is the bilinear map* $[\cdot, \cdot] : \mathbf{A} \times \mathbf{A} \to \mathbf{A}$ *defined by:*

$$[a, b] := ab - (-1)^{\tilde{a}\tilde{b}} ba$$

for all homogeneous elements $a, b \in \mathbf{A}$. *The algebra* $\mathbf{A}$ *is called* **supercommutative** *if* $[a, b] = 0$ *for all* $a, b \in \mathbf{A}$. *A* $\mathbb{Z}_2$*-graded supercommutative associative (real) algebra* $A = A_0 + A_1$ *will simply be called a* **superalgebra**.

Example 1: The exterior algebra $\wedge E = \wedge^{even} E + \wedge^{odd} E$ over a finite dimensional vector space E is a superalgebra.

DEFINITION 16. *A* **super Lie bracket** *on a* $\mathbb{Z}_2$*-graded vector space* $V = V_0 + V_1$ *is a bilinear map* $[\cdot, \cdot] : V \times V \to V$ *such that for all* $x, y, z \in V_0 \cup V_1$ *we have:*

i) $\widetilde{[x, y]} = \tilde{x} + \tilde{y}$,
ii) $[x, y] = -(-1)^{\tilde{x}\tilde{y}} [y, x]$ *and*
iii) $[x, [y, z]] = [[x, y], z] + (-1)^{\tilde{x}\tilde{y}} [y, [x, z]]$ *("super Jacobi identity").*

The $\mathbb{Z}_2$*-graded algebra with underlying* $\mathbb{Z}_2$*-graded vector space* V *and product defined by the super Lie bracket* $[\cdot, \cdot]$ *is called a* **super Lie algebra**.

Example 2: The supercommutator of any associative $\mathbb{Z}_2$-graded algebra $\mathbf{A}$ is a super Lie bracket and hence defines on it the structure of super Lie algebra. For example, we may take $\mathbf{A} = \mathrm{End}(V)$ with the obvious structure of $\mathbb{Z}_2$-graded associative algebra (with unit). The corresponding super Lie algebra is denoted by $\mathfrak{gl}(V)$ and is called the **general linear super Lie algebra**.

Let M_0 be a (differentiable) manifold of dimension m. We denote by $\mathcal{C}^\infty_{M_0}$ its sheaf of functions. Sections of the sheaf $\mathcal{C}^\infty_{M_0}$ over an open set $U \subset M_0$ are simply smooth functions on U: $\mathcal{C}^\infty_{M_0}(U) = C^\infty(U)$. Now let $\mathcal{A} = \mathcal{A}_0 + \mathcal{A}_1$ be a sheaf of superalgebras over M_0.

DEFINITION 17. *The pair* $M = (M_0, \mathcal{A})$ *is called a (differentiable)* **supermanifold** *of* **dimension** $\dim M = m|n$ *over* M_0 *if for all* $p \in M_0$ *there exists an open neighborhood* $U \ni p$ *and a rank* n *free sheaf* $\mathcal{E}_U$ *of* $\mathcal{C}^\infty_U$*-modules over* U *such that* $\mathcal{A}|_U \cong \wedge \mathcal{E}_U$ *(as sheaves of superalgebras). A* **function** *on* M *(over an open set* $U \subset M_0$*) is by definition a section of* $\mathcal{A}$ *(over* U*). The sheaf* $\mathcal{A} = \mathcal{A}_M$ *is called the* **sheaf of functions** *on* M *and* M_0 *is called the manifold* **underlying** *the supermanifold* M. *Let* $M = (M_0, \mathcal{A}_M)$ *and* $N = (N_0, \mathcal{A}_N)$ *be supermanifolds. A*

morphism *$\varphi : M \to N$ is a pair $\varphi = (\varphi_0, \varphi^*)$, where $\varphi_0 : M_0 \to N_0$ is a smooth map and $\varphi^* : \mathcal{A}_N \to (\varphi_0)_*\mathcal{A}_M$ is a morphism of sheaves of superalgebras. It is called an* **isomorphism** *if φ_0 is a diffeomorphism and φ^* an isomorphism. An isomorphism $\varphi : M \to M$ is called an* **automorphism** *of M. The set of all morphisms $\varphi : M \to N$ (respectively, automorphisms $\varphi : M \to M$) is denoted by* $\mathrm{Mor}(M, N)$ *(respectively,* $\mathrm{Aut}(M)$*).*

From Def. 17 it follows that there exists a canonical epimorphism of sheaves $\epsilon^* : \mathcal{A} \to \mathcal{C}^\infty_{M_0}$, which is called the **evaluation map**. Its kernel is the ideal generated by $\mathcal{A}_1$: $\ker \epsilon^* = \langle \mathcal{A}_1 \rangle = \mathcal{A}_1 + \mathcal{A}_1^2$.

Given supermanifolds L, M, N and morphisms $\psi \in \mathrm{Mor}(L, M)$ and $\varphi \in \mathrm{Mor}(M, N)$, there is a **composition** $\varphi \circ \psi \in \mathrm{Mor}(L, N)$ defined by:

$$(\varphi \circ \psi)_0 = \varphi_0 \circ \psi_0 \quad \text{and} \quad (\varphi \circ \psi)^* = \psi^* \circ \varphi^* .$$

Here we have used the same symbol ψ^* for the map $(\varphi_0)_*\mathcal{A}_M \to (\varphi_0)_*(\psi_0)_*\mathcal{A}_L = (\varphi_0 \circ \psi_0)_*\mathcal{A}_L$ induced by $\psi^* : \mathcal{A}_M \to (\psi_0)_*\mathcal{A}_L$. Similarly, if $\varphi : M \to N$ is an isomorphism, then we can define it **inverse** isomorphism by:

$$\varphi^{-1} := (\varphi_0^{-1}, (\varphi^*)^{-1}) : N \to M .$$

Here, again, we have used the same notation $(\varphi^*)^{-1}$ for the map $\mathcal{A}_M \to (\varphi_0^{-1})_*\mathcal{A}_N$ induced by $(\varphi^*)^{-1} : (\varphi_0)_*\mathcal{A}_M \to \mathcal{A}_N$. Finally, for every supermanifold $M = (M_0, \mathcal{A})$, there is the **identity** automorphism $\mathrm{Id}_M := (\mathrm{Id}_{M_0}, \mathrm{Id}_{\mathcal{A}})$. The above operations turn the set $\mathrm{Aut}(M)$ into a group.

Example 3: Let $E \to M_0$ be a (smooth) vector bundle of rank n over the m-dimensional manifold M_0 and $\mathcal{E}$ its sheaf of sections. It is a locally free sheaf of $\mathcal{C}^\infty_{M_0}$-modules and $SM(E) := (M_0, \wedge\mathcal{E})$ is a supermanifold of dimension $m|n$. Its evaluation map is the canonical projection onto 0-forms $\wedge\mathcal{E} = \mathcal{C}^\infty_{M_0} + \sum_{j=1}^n \wedge^j \mathcal{E} \to \mathcal{C}^\infty_{M_0}$. It is well known, see [**Ba**], that any supermanifold is isomorphic to a supermanifold of the form $SM(E)$. However, the isomorphism is not canonical, unless $n = 0$.

Example 4: Any manifold $(M_0, \mathcal{C}^\infty_{M_0})$ of dimension m can be considered as a supermanifold of dimension $m|0$. In fact, it is associated to the vector bundle of rank 0 over M_0 via the construction of Example 3. For any supermanifold $M = (M_0, \mathcal{A})$ the pair $(\mathrm{Id}_{M_0}, \epsilon^*)$ defines a canonical morphism $\epsilon : M_0 \to M$. The composition of ϵ with the canonical constant map $p : \{p\} \to M_0$ ($p \in M_0$) defines a morphism $\epsilon_p = (p, \epsilon_p^*) : \{p\} \to M$. The epimorphism $\epsilon_p^* : \mathcal{A} \to \mathbb{R}$ onto the constant sheaf is called the **evaluation at** p:

$$\epsilon_p^* f = (\epsilon^* f)(p) .$$

$f(p) := \epsilon_p^* f \in \mathbb{R}$ is called the **value** of f **at** the point p.

Example 5: Let $V = V_0 + V_1$ be a $\mathbb{Z}_2$-graded vector space of dimension $m|n$ and $E_V = V_1 \times V_0 \to V_0$ the trivial vector bundle over V_0 with fibre V_1. Then to V we can canonically associate the supermanifold $SM(V) := SM(E_V)$, see Example 3.

Let $(x_0^i) = (x_0^1, \dots, x_0^m)$ be local coordinates for M_0 defined on an open set $U \subset M_0$, $\mathcal{E}_U$ a rank n free sheaf of $\mathcal{C}^\infty_U$-modules over U and $\phi : \wedge\mathcal{E}_U \to \mathcal{A}|_U$ an isomorphism. We can choose sections $(\xi_0^j) = (\xi_0^1, \dots, \xi_0^n)$ of $\mathcal{E}_U$ which generate $\mathcal{E}_U$ freely over $\mathcal{C}^\infty_U$. Then any section of $\wedge\mathcal{E}_U$ is of the form:

$$f = \sum_{\alpha \in \mathbb{Z}_2^n} f_\alpha(x_0^1, \dots, x_0^m)\xi_0^\alpha , \quad f_\alpha(x_0^1, \dots, x_0^m) \in C^\infty(U) , \tag{21}$$

where $\xi_0^\alpha := (\xi_0^1)^{\alpha_1} \wedge \ldots \wedge (\xi_0^n)^{\alpha_n}$ for $\alpha = (\alpha_1, \ldots, \alpha_n)$. The tuple (ϕ, x_0^i, ξ_0^j) is called a **local coordinate system** for M over U. The open set $U \subset M_0$ is called a **coordinate neighborhood** for M. Any function on M over U is of the form $\phi(f)$ for some section f as in (21). The functions $x^i := \phi(x_0^i) \in \mathcal{A}(U)_0$, $\xi^j := \phi(\xi_0^j) \in \mathcal{A}_1$ are called **local coordinates** for M over U. The evaluation map $\epsilon^* : \mathcal{A} \to \mathcal{C}^\infty_{M_0}$ is expressed in a local coordinate system simply by putting $\xi_0^1 = \cdots = \xi_0^n = 0$ in (21):

$$\epsilon^*(\phi(f)) = f(x_0^1, \ldots, x_0^m, 0, \ldots, 0) := f_{(0,\ldots,0)}(x_0^1, \ldots, x_0^m).$$

In particular, we have $\epsilon^* x^i = x_0^i$ and $\epsilon^* \xi^j = 0$.
Example 6: Let $V = V_0 + V_1$ be a $\mathbb{Z}_2$-graded vector space and $(x^i, \xi^j) = (x^i, \ldots, x^m, \xi^1, \ldots, \xi^n)$ a basis of the $\mathbb{Z}_2$-graded vector space $V^* = \mathrm{Hom}(V, \mathbb{R})$. Then (x^i, ξ^j) can be considered as global coordinates on the supermanifold $SM(V)$, i.e. as local coordinates for $SM(V)$ over $V_0 = SM(V)_0$, see Example 5.

Let M and N be supermanifolds of dimension $m|n$ and $p|q$ respectively. In local coordinates (x^i, ξ^j) for M and (y^k, η^l) for N a morphism φ is expressed by p even functions $y^k(x^1, \ldots, x^m, \xi^1, \ldots, \xi^n) := \varphi^* y^k$ and q odd functions $\eta^l(x^1, \ldots, x^m, \xi^1, \ldots, \xi^n) := \varphi^* \eta^l$.

There exists a supermanifold $M \times N = (M_0 \times N_0, \mathcal{A}_{M\times N})$ called the **product** of the supermanifolds M and N and morphisms $\pi_M : M \times N \to M$, $\pi_N : M \times N \to N$ such that $(\pi_M^* x^i, \pi_N^* y^k, \pi_M^* \xi^j, \pi_N^* \eta^l)$ are local coordinates for $M \times N$ over $U \times V$ if (x^i, ξ^j) are local coordinates for M over U and (y^k, η^l) are local coordinates for N over V. The morphism $\pi_1 = \pi_M$ (respectively, $\pi_2 = \pi_N$) is called the **projection** of $M \times N$ onto the first (respectively, second) factor. Given morphisms $\varphi_i : M_i \to N_i$, $i = 1, 2$, there is a corresponding morphism $\varphi_1 \times \varphi_2 : M_1 \times M_2 \to N_1 \times N_2$ such that $\pi_{N_1} \circ (\varphi_1 \times \varphi_2) = \varphi_1 \circ \pi_{M_1}$ and $\pi_{N_2} \circ (\varphi_1 \times \varphi_2) = \varphi_2 \circ \pi_{M_2}$.

As next, we will discuss the notion for tangency on supermanifolds. For this purpose, we recall the following definition.

DEFINITION 18. *An endomorphism* $X = X_0 + X_1 \in \mathrm{End}(\mathbf{A}) = \mathrm{End}_{\mathbb{R}}(\mathbf{A})$ *(here* $\tilde{X}_\alpha = \alpha$, $\alpha = 0, 1$*) of a* $\mathbb{Z}_2$*-graded algebra* $\mathbf{A}$ *is called a* **derivation** *if it satisfies the Leibniz-rule*

$$X_\alpha(ab) = X_\alpha(a)b + (-1)^{\alpha\tilde{a}} a X_\alpha(b)$$

for all homogeneous $a, b \in \mathbf{A}$ *and* $\alpha \in \mathbb{Z}_2$. *The* $\mathbb{Z}_2$*-graded vector space of all derivations of* $\mathbf{A}$ *is denoted by* $\mathrm{Der}\,\mathbf{A} = (\mathrm{Der}\,\mathbf{A})_0 + (\mathrm{Der}\,\mathbf{A})_1$.

Notice that the supercommutator on $\mathrm{End}(\mathbf{A})$ restricts to a super Lie bracket on $\mathrm{Der}\,\mathbf{A}$.

DEFINITION 19. *Let* $M = (M_0, \mathcal{A})$ *be a supermanifold. The* **tangent sheaf** *of* M *is the sheaf of derivations of* $\mathcal{A}$ *and is denoted by* $\mathcal{T}_M = (\mathcal{T}_M)_0 + (\mathcal{T}_M)_1$. *A* **vector field** *on* M *is a section of* $\mathcal{T}_M$. *The* **cotangent sheaf** *is the sheaf* $\mathcal{T}_M^* = \mathrm{Hom}_{\mathcal{A}}(\mathcal{T}_M, \mathcal{A})$. *The* **full tensor superalgebra** *over* $\mathcal{T}_M$ *is the sheaf of superalgebras generated by tensor products (*$\mathbb{Z}_2$*-graded over* $\mathcal{A}$*) of* $\mathcal{T}_M$ *and* $\mathcal{T}_M^*$. *It is denoted by* $\otimes_{\mathcal{A}}\langle \mathcal{T}_M, \mathcal{T}_M^* \rangle$. *A* **tensor field** *on* M *is a section of* $\otimes_{\mathcal{A}}\langle \mathcal{T}_M, \mathcal{T}_M^* \rangle$.

Explicitly, a section $X \in \mathcal{T}_M(U)$ ($U \subset M_0$ open) associates to any open subset $V \subset U$ a derivation $X|_V \in \mathrm{Der}\,\mathcal{A}(V)$ such that

$$X|_V(f|_V) = X|_U(f)|_V \quad \text{for all} \quad f \in \mathcal{A}(U).$$

Given local coordinates (x^i, ξ^j) on M over U there exist unique vector fields

$$\frac{\partial}{\partial x^i} \in \mathcal{T}_M(U)_0 \quad \text{and} \quad \frac{\partial}{\partial \xi^j} \in \mathcal{T}_M(U)_1$$

such that

$$\frac{\partial x^k}{\partial x^i} = \delta_i^k\,, \quad \frac{\partial \xi^l}{\partial x^i} = 0\,, \quad \frac{\partial x^k}{\partial \xi^j} = 0\,, \quad \frac{\partial \xi^l}{\partial \xi^j} = \delta_j^l\,.$$

(Here, of course, $\delta_i^k \in \mathcal{A}(U)$ is the unit element in the algebra $\mathcal{A}(U)$ if $i = k$ and is zero otherwise.) Moreover, the vector fields $(\partial/\partial x^i, \partial/\partial \xi^j)$ freely generate $\mathcal{T}_M(U) \cong \mathrm{Der}\,\mathcal{A}(U)$ over $\mathcal{A}(U)$. This shows that $\mathcal{T}_M$ is a locally free sheaf of rank $m|n = \dim M$ over $\mathcal{A}$. It is also a sheaf of super Lie algebras; simply because $\mathrm{Der}\,\mathcal{A}(U)$ is a subalgebra of the super Lie algebra $\mathrm{End}_{\mathbb{R}}\,\mathcal{A}(U)$ for all open $U \subset M_0$. As in the case of ordinary manifolds, given a vector field X there exists a unique derivation $\mathcal{L}_X$ of the full tensor superalgebra $\otimes_{\mathcal{A}}\langle \mathcal{T}_M, \mathcal{T}_M^*\rangle$ over $\mathcal{T}_M$ compatible with contractions such that

$$\mathcal{L}_X f = X(f) \quad \text{and} \quad \mathcal{L}_X Y = [X, Y]$$

for all functions f and vector fields Y on M ($[X, Y]$ is the *super*commutator of vector fields).

DEFINITION 20. *Let $M = (M_0, \mathcal{A})$ be a supermanifold. A* **tangent vector** *to M at $p \in M_0$ is an $\mathbb{R}$-linear map $v = v_0 + v_1 : \mathcal{A}_p \to \mathbb{R}$ such that*

$$v_\alpha(fg) = v_\alpha(f)\epsilon_p^*(g) + (-1)^{\alpha\tilde{f}}\epsilon_p^*(f)v_\alpha(g)\,, \quad \alpha = 0, 1\,,$$

for all germs of functions $f, g \in \mathcal{A}_p$ of pure degree. ($\mathcal{A}_p$ denotes the stalk of $\mathcal{A}$ at p.) The $\mathbb{Z}_2$-graded vector space of all tangent vectors to M at p is denoted by $T_pM = (T_pM)_0 + (T_pM)_1$ and is called the **tangent space** *to M at p.*

Let X be a vector field defined on some open set $U \subset M_0$ and $p \in U$. Then we can define the **value** $X(p) \in T_pM$ of X **at** p:

$$X(p)(f) := \epsilon_p^*(X(f))\,, \quad f \in \mathcal{A}_p\,.$$

However, unless $\dim M = m|n = m|0$, a vector field is not determined by its values at all $p \in M_0$. The above definition of value at a point p is naturally extended to arbitrary tensor fields S; the value of S at p is denoted by $S(p)$. Again, unless $\dim M = m|n = m|0$, a tensor field on M is not determined by its values at all $p \in M_0$.

Given a morphism $\varphi : M \to N$, to any local vector field $X \in \mathcal{T}_M(U)$ on M we can associate a vector field $d\varphi X \in (\varphi^*\mathcal{T}_N)(U)$ on N with values in $\mathcal{A}_M$ which is defined by:

$$(d\varphi X)(f) = X(\varphi^* f)\,, \quad f \in \mathcal{A}_N(V)\,,$$

where $V \subset N_0$ is an open set such that $\varphi_0^{-1}(V) \supset U$. We recall that $\varphi^*\mathcal{T}_N$ is the sheaf of $\mathcal{A}_M$-modules over M_0 defined by:

$$\varphi^*\mathcal{T}_N := \mathcal{A}_M \otimes_{\varphi_0^{-1}\mathcal{A}_N} \varphi_0^{-1}\mathcal{T}_N\,.$$

Here the action of $\varphi_0^{-1}\mathcal{A}_N$ on $\mathcal{A}_M$ is defined by the map

$$\varphi_0^{-1}\mathcal{A}_N \to \varphi_0^{-1}\varphi_*\mathcal{A}_M \to \mathcal{A}_M$$

induced by $\varphi^* : \mathcal{A}_N \to \varphi_* \mathcal{A}_M$. By the above construction, we obtain a section $d\varphi$ of the sheaf $\mathrm{Hom}_{\mathcal{A}}(\mathcal{T}_M, \varphi^* \mathcal{T}_N)$, which is expressed with respect to local coordinates $(u^1, \ldots, u^{m+n}) = (x^1, \ldots, x^m, \xi^1, \ldots, \xi^n)$ on M and $(v^1, \ldots, v^{p+q}) = (y^1, \ldots, y^p, \eta^1, \ldots, \eta^q)$ on N by the Jacobian matrix $(\frac{\partial \varphi^* v^i}{\partial u^j})$. The value $d\varphi(p) \in \mathrm{Hom}(T_pM, T_{\varphi_0(p)}N)$ of the differential at a point $p \in M_0$ is defined by:

$$d\varphi(p)X(p) = (d\varphi X)(p)$$

for all vector fields X on M. The differential $d\epsilon$ of the canonical morphism $\epsilon : M_0 \to M$ provides the canonical isomorphism $d\epsilon(p) : T_pM_0 \stackrel{\sim}{\to} (T_pM)_0$ for all $p \in M$.

DEFINITION 21. *A morphism $\varphi : M \to N$ is called an* **immersion** *(respectively, a* **submersion***) if $d\varphi$ has constant rank $m|n = \dim M$ (respectively, $p|q = \dim N$), i.e. if for all local coordinates as above the matrix $(\epsilon^*(\partial \varphi^* y^i / \partial x^j))$ has constant rank m (respectively, p) and $(\epsilon^*(\partial \varphi^* \eta^i / \partial \xi^j))$ has constant rank n (respectively, q). An immersion $\varphi : M \to N$ is called* **injective** *(respectively, an* **embedding**, *a* **closed embedding***) and is denoted by $\varphi : M \hookrightarrow N$ if $\varphi_0 : M_0 \to N_0$ is injective (respectively, an embedding, a closed embedding). Two immersions $\varphi : M \to N$ and $\varphi' : M' \to N$ are called* **equivalent** *if there exists an isomorphism $\psi : M \to M'$ such that $\varphi = \varphi' \circ \psi$. A* **submanifold** *(respectively, an* **embedded submanifold**, *a* **closed submanifold***) is an equivalence class of injective immersions (respectively, embeddings, closed embeddings).*

Notice that for any supermanifold the canonical morphism $\epsilon = (\mathrm{Id}_{M_0}, \epsilon^*) : M_0 \hookrightarrow M$ is a closed embedding.

As for ordinary manifolds, immersions and submersions admit adapted coordinates:

PROPOSITION 15. *(see* **[L]**, **[K]***) Let M and N be supermanifolds of dimension $\dim M = m|n$ and $\dim N = p|q$. A morphism $\varphi : M \to N$ is an immersion (respectively, submersion) if and only if for all $(x, y = \varphi_0(x)) \in M_0 \times \varphi_0(M_0) \subset M_0 \times N_0$ there exists local coordinates $(x^1, \ldots, x^m, \xi^1, \ldots, \xi^n)$ for M defined near x and $(y^1, \ldots, y^p, \eta^1, \ldots, \eta^q)$ for N defined near y such that*

$$\varphi^* y^i = \begin{cases} x^i & \text{for} \quad i = 1, \ldots, m \leq p \\ 0 & \text{for} \quad m < i \leq p \end{cases} \quad \text{and} \quad \varphi^* \eta^j = \begin{cases} \xi^j & \text{for} \quad j = 1, \ldots, n \leq q \\ 0 & \text{for} \quad n < j \leq q \end{cases}$$

(respectively,

$$\varphi^* y^i = x^i \quad \text{for} \quad i = 1, \ldots, p \leq m \quad \text{and} \quad \varphi^* \eta^j = \xi^j \quad \text{for} \quad j = 1, \ldots, q \leq n).$$

For any submanifold $\varphi : M \hookrightarrow N$ we define its **vanishing ideal** $\mathcal{J}_y \subset (\mathcal{A}_N)_y$ **at** $y \in N_0$ as follows: $\mathcal{J}_y := (\mathcal{A}_N)_y$ if $y \notin \varphi_0(M_0)$ and $\mathcal{J}_y := \ker(r \circ \varphi^*)$ if $y = \varphi_0(x)$, $x \in M_0$, where $r : ((\varphi_0)_* \mathcal{A}_M)_y \to (\mathcal{A}_M)_x$ is the natural restriction homomorphism. The union $\mathcal{J} := \cup_{y \in N_0} \mathcal{J}_y \subset \mathcal{A}_N$ is called the **vanishing ideal** of the submanifold $\varphi : M \hookrightarrow N$. By Prop. 15, it has the following property (P): *If $y \in \varphi_0(M_0)$ then there exists $p-m$ even functions (y^i) and $q-n$ odd functions η^j on N vanishing at y whose germs at y generate $\mathcal{J}_y$ and which can be complemented to local coordinates for N defined near y. (Here $m|n = \dim M$ and $p|q = \dim N$.) We denote by $Z_U \subset N_0$ the submanifold defined by the equations $\epsilon^* y^i = 0$ on some sufficiently small open neighborhood $U \subset N_0$ of y. Then U can be chosen such that the germs of the functions (y^i, η^j) at z generate $\mathcal{J}_z$ for all $z \in Z_U$.*

Conversely, let $\varphi_0 : M_0 \hookrightarrow N_0$ be any submanifold and $\mathcal{J}_y \subset (\mathcal{A}_N)_y$, $y \in N$, a collection of ideals with the above property (P) and such that $J_y = (\mathcal{A}_N)_y$ for

$y \notin \varphi_0(M_0)$, then there exists a supermanifold $M = (M_0, \mathcal{A}_M)$ and an injective immersion $\varphi = (\varphi_0, \varphi^*) : M \hookrightarrow N$ with $\mathcal{J} = \cup_{y \in N} \mathcal{J}_y$ as its vanishing ideal.

Notice that if $M \hookrightarrow N$ is a *closed* submanifold then its vanishing ideal can be defined directly as the sheaf of ideals $\mathcal{J} := \ker \varphi^*$.

4.5. Pseudo-Riemannian metrics, connections and quaternionic structures on supermanifolds. Let $\mathbf{A}$ be a superalgebra and T a free $\mathbf{A}$-module of rank $m|n$.

DEFINITION 22. *An even (respectively, odd)* **bilinear form** *on T is a biadditive map $g : T \times T \to A$ such that*

$$g(aX, bY) = (-1)^{\tilde{b}\tilde{X}} abg(X, Y)$$

$$(\textit{respectively,} \quad g(aX, bY) = (-1)^{\tilde{b}\tilde{X} + \tilde{a} + \tilde{b}} abg(X, Y)),$$

for all homogeneous $a, b \in \mathbf{A}$ and $X, Y \in T$. A bilinear form g on T is called **supersymmetric** *(respectively,* **super skew symmetric***) if*

$$g(X, Y) = (-1)^{\tilde{X}\tilde{Y}} g(Y, X)$$

$$(\textit{respectively,} \quad g(X, Y) = -(-1)^{\tilde{X}\tilde{Y}} g(Y, X)),$$

for all homogeneous $X, Y \in T$. It is called **nondegenerate** *if*

$$T \ni X \mapsto g(X, \cdot) \in T^* = \mathrm{Hom}_A(T, \mathbf{A})$$

is an isomorphism of $\mathbf{A}$-modules. The $\mathbf{A}$-module of bilinear forms on T is denoted by $\mathrm{Bil}_{\mathbf{A}}(T) = \mathrm{Bil}_{\mathbf{A}}(T)_0 + \mathrm{Bil}_{\mathbf{A}}(T)_1$.

Notice that $\mathrm{Bil}_{\mathbf{A}}(T) \cong \mathrm{Hom}_{\mathbf{A}}(T, T^*) \cong T^* \otimes_{\mathbf{A}} T^*$, where $T^* = \mathrm{Hom}_{\mathbf{A}}(T, \mathbf{A})$.

For a supermanifold $M = (M_0, \mathcal{A})$ the sheaf $\mathrm{Bil}_{\mathcal{A}} \mathcal{T}_M$ of bilinear forms on $\mathcal{T}_M$ is defined in the obvious way such that $(\mathrm{Bil}_{\mathcal{A}} \mathcal{T}_M)(U) = \mathrm{Bil}_{\mathcal{A}(U)}(\mathcal{T}_M(U))$ for every open subset $U \subset M_0$ with the property that $\mathcal{T}_M(U)$ is a free $\mathcal{A}(U)$-module. We have obvious isomorphisms of sheaves of $\mathcal{A}$-modules: $\mathrm{Bil}_{\mathcal{A}} \mathcal{T}_M \cong \mathrm{Hom}_{\mathcal{A}}(\mathcal{T}_M, \mathcal{T}_M^*) \cong \mathcal{T}_M^* \otimes_{\mathcal{A}} \mathcal{T}_M^*$. Since any section g of $\mathrm{Bil}_{\mathcal{A}} \mathcal{T}_M$ can be considered as a tensor field on M, it has a well defined value $g(p) \in \mathrm{Bil}_{\mathbb{R}}(T_pM)$ for all $p \in M_0$. The restriction $g(p)|(T_pM)_0 \times (T_pM)_0$ defines a section g_0 of $\mathrm{Bil}_{C^\infty_{M_0}} \mathcal{T}_{M_0}$ via the canonical identification $d\epsilon(p) : T_pM_0 \xrightarrow{\sim} (T_pM)_0$.

DEFINITION 23. *A* **pseudo-Riemannian metric** *on a supermanifold $M = (M_0, \mathcal{A})$, M_0 connected, is an even nondegenerate supersymmetric section g of $\mathrm{Bil}_{\mathcal{A}} \mathcal{T}_M$. The* **signature** *$(k, l)$ of g is the signature of the pseudo-Riemannian metric g_0 on M_0. The pseudo-Riemannian metric g is said to be a* **Riemannian metric** *if g_0 is Riemannian. Let $M = (M, \mathcal{A})$ be a supermanifold and $\mathcal{E}$ a locally free sheaf of $\mathcal{A}$-modules. A* **connection** *on $\mathcal{E}$ is an even section ∇ of the sheaf $\mathrm{Hom}_{\mathcal{A}}(\mathcal{T}_M, \mathrm{End}_{\mathbb{R}} \mathcal{E})$, which to any vector field X on M associates a section ∇_X of $\mathrm{End}_{\mathbb{R}} \mathcal{E}$ such that*

$$\nabla_X fs = X(f)s + (-1)^{\tilde{X}\tilde{f}} f \nabla_X s$$

for all vector fields X on M, functions f on M and sections s of $\mathcal{E}$ of pure degree. The **curvature** *R of ∇ is the even super skew symmetric section of $\mathrm{End}_{\mathcal{A}} \mathcal{E} \otimes_{\mathcal{A}} \mathrm{Bil}_{\mathcal{A}} \mathcal{T}_M$ defined by:*

$$R(X, Y) := \nabla_X \nabla_Y - (-1)^{\tilde{X}\tilde{Y}} \nabla_Y \nabla_X - \nabla_{[X,Y]}$$

for all vector fields X, Y on M of pure degree. A **connection on a supermanifold** *M is by definition a connection on its tangent sheaf $\mathcal{T}_M$. Its* **torsion** *is the even super skew symmetric section of $\mathcal{T}_M \otimes_{\mathcal{A}} \mathrm{Bil}_{\mathcal{A}} \mathcal{T}_M$ defined by:*

$$T(X,Y) := \nabla_X Y - (-1)^{\tilde{X}\tilde{Y}} \nabla_Y X - [X,Y]$$

for all vector fields X, Y on M of pure degree.

As for ordinary manifolds, a connection ∇ on a supermanifold M induces a connection on the full tensor superalgebra $\otimes_{\mathcal{A}}\langle \mathcal{T}_M, \mathcal{T}_M^* \rangle$. In particular, if g is e.g. an even section of $\mathrm{Bil}_{\mathcal{A}} \mathcal{T}_M \cong \mathcal{T}_M^* \otimes \mathcal{T}_M^*$ then we have

$$(\nabla_X g)(Y,Z) = Xg(Y,Z) - g(\nabla_X Y, Z) - (-1)^{\tilde{X}\tilde{Y}} g(Y, \nabla_X Z)$$

for all vector fields X, Y and Z on M of pure degree. As in the case of ordinary manifolds, see e.g. [**O'N**], one can prove that a pseudo-Riemannian supermanifold (M,g) has a unique torsionfree connection $\nabla = \nabla^g$ such that $\nabla g = 0$. We will call this connection the **Levi-Civita connection** of (M,g). It is computable from the following superversion of the Koszul-formula:

$$\begin{aligned} 2g(\nabla_X Y, Z) &= Xg(Y,Z) + (-1)^{\tilde{X}(\tilde{Y}+\tilde{Z})} Y g(Z,X) \\ &\quad - (-1)^{\tilde{Z}(\tilde{X}+\tilde{Y})} Z g(X,Y) - g(X,[Y,Z]) \\ &\quad + (-1)^{\tilde{X}(\tilde{Y}+\tilde{Z})} g(Y,[Z,X]) + (-1)^{\tilde{Z}(\tilde{X}+\tilde{Y})} g(Z,[X,Y]) \end{aligned} \tag{22}$$

for all vector fields X, Y and Z on M of pure degree.

Next we are going to define quaternionic supermanifolds and quaternionic Kähler supermanifolds. First we define the notion of almost quaternionic structure.

DEFINITION 24. *Let $M = (M_0, \mathcal{A})$ be a supermanifold. An* **almost complex structure** *on M is an even global section $J \in (\mathrm{End}_{\mathcal{A}} \mathcal{T}_M)(M_0)$ such that $J^2 = -\mathrm{Id}$. An* **almost hypercomplex** *structure on M is a triple $(J_\alpha) = (J_1, J_2, J_3)$ of almost complex structures on M satisfying $J_1 J_2 = J_3$. An* **almost quaternionic structure** *on M is a subsheaf $Q \subset \mathrm{End}_{\mathcal{A}} \mathcal{T}_M$ with the following property: for every $p \in M_0$ there exists an open neighborhood $U \subset M_0$ and an almost hypercomplex structure (J_α) on $M|_U$ such that $Q(U)$ is a free $\mathcal{A}(U)$-module of rank $3|0$ with basis (J_1, J_2, J_3). A pair (M, J) (respectively, $(M, (J_\alpha))$, (M, Q)) as above is called an* **almost complex supermanifold** *(respectively,* **almost hypercomplex supermanifold**, **almost quaternionic supermanifold***).*

Second we introduce the basic compatibility conditions between almost quaternionic structures, connections and pseudo-Riemannian metrics.

DEFINITION 25. *Let (M,Q) be an almost quaternionic supermanifold of dimension $\dim M = m|n$. A connection ∇ on (M,Q) is called an* **almost quaternionic connection** *if ∇ preserves Q, i.e. if $\nabla_X S$ is a section of Q for any vector field X on M and any section S of Q. A* **quaternionic connection** *on (M,Q) is a torsion-free almost quaternionic connection. If the almost quaternionic structure Q on M admits a quaternionic connection, then it is called* **1-integrable** *or* **quaternionic structure***. In this case the pair (M,Q) is called a* **quaternionic supermanifold,** *provided that $m = \dim M_0 > 4$.*

DEFINITION 26. *A (pseudo-) Riemannian metric g on an almost quaternionic supermanifold (M,Q) is called* **Hermitian** *if sections of Q are g-skew symmetric,* *i.e. if* $g(SX,Y) = -g(X,SY)$ *for all sections* S *of* Q *and vector fields* X, Y *on* M. *In this case the triple* (M,Q,g) *is called an* **almost quaternionic** *(pseudo-)* **Hermitian supermanifold**. *If, moreover, the Levi-Civita connection* ∇^g *of the* Q*-Hermitian metric* g *is quaternionic, then* (M,Q,g) *is called a* **quaternionic** *(pseudo-)* **Kähler supermanifold**, *provided that* $m = \dim M_0 > 4$.

We recall that to any (pseudo-) Riemannian metric g on a supermanifold M we have associated the (pseudo-) Riemannian metric g_0 on the manifold M_0, see Def. 23. Now we will associate an almost quaternionic structure Q_0 on M_0 to any almost quaternionic structure Q on M. To any even section S of $\mathrm{End}_{\mathcal{A}} \mathcal{T}_M$ we associate a section S_0 of $\mathrm{End}_{\mathcal{C}^\infty_{M_0}} \mathcal{T}_{M_0}$ defined by $S_0(p) = S(p)|T_pM_0$, where $S(p) \in (\mathrm{End}_{\mathbb{R}}(T_pM))_0$ is the value of the tensor field S at $p \in M_0$ and, as usual, T_pM_0 is canonically identified with $(T_pM)_0$. Let $Q_0 \subset \mathrm{End}_{\mathcal{C}^\infty_{M_0}} \mathcal{T}_{M_0}$ be the subsheaf of $\mathcal{C}^\infty_{M_0}$-modules spanned by the local sections of the form S_0, where S is a local section of Q. Then Q_0 is a locally free sheaf of rank 3 and hence it defines a rank 3 subbundle of the vectorbundle $\mathrm{End}\, TM_0$, more precisely, an almost quaternionic structure on M_0 in the usual sense, see Def. 5.

Finally we give the definition of quaternionic supermanifolds and of quaternionic Kähler supermanifolds of dimension $4|n$.

DEFINITION 27. *An almost quaternionic supermanifold* (M,Q) *of dimension* $\dim M = 4|n$ *is called a* **quaternionic supermanifold** *if*

i) *there exists a quaternionic connection* ∇ *on* (M,Q),
ii) (M_0,Q_0) *is a quaternionic manifold in the sense of Def. 8.*

An almost quaternionic Hermitian supermanifold (M,Q,g) *of dimension* $\dim M = 4|n$ *is called a* **quaternionic Kähler supermanifold** *if*

i) *the Levi-Civita connection* ∇^g *preserves* g *and*
ii) (M_0,Q_0) *is a quaternionic Kähler manifold in the sense of Def. 11.*

4.6. Supergroups. Let $V = V_0 + V_1$ be a $\mathbb{Z}_2$-graded vector space. We recall that $\mathfrak{gl}(V)$ denotes the general linear super Lie algebra. Its super Lie bracket is the supercommutator on $\mathrm{End}_{\mathbb{R}} V$, see Example 2.

DEFINITION 28. *A* **representation** *of a super Lie algebra* $\mathfrak{g}$ *on a* $\mathbb{Z}_2$*-graded vector space* V *is a homomorphism of super Lie algebras* $\mathfrak{g} \to \mathfrak{gl}(V)$. *The* **adjoint representation** *of* $\mathfrak{g}$ *is the representation* $\mathrm{ad} : \mathfrak{g} \ni x \mapsto \mathrm{ad}_x \in \mathfrak{gl}(\mathfrak{g})$ *defined by:*

$$\mathrm{ad}_x y = [x,y]\,, \quad x,y \in \mathfrak{g}\,.$$

Here $[\cdot,\cdot]$ *denotes the super Lie bracket in* $\mathfrak{g}$. *A* **linear super Lie algebra** *is a subalgebra* $\mathfrak{g} \subset \mathfrak{gl}(V)$.

Notice that, by the super Jacobi identity, see Def. 16, ad_x is a derivation of $\mathfrak{g}$ for all $x \in \mathfrak{g}$.

Now let E be a module over a superalgebra $\mathbf{A}$; say $E = V \otimes \mathbf{A}$, where V is a $\mathbb{Z}_2$-graded vector space (and $\otimes$ stands for the $\mathbb{Z}_2$-graded tensor product over $\mathbb{R}$). The invertible elements of $(\mathrm{End}_{\mathbf{A}} E)_0$ form a group, which is denoted by $\mathrm{GL}_{\mathbf{A}}(E)$. Given a $\mathbb{Z}_2$-graded vector space V, the correspondence

$$\mathbf{A} \mapsto \mathrm{GL}_{\mathbf{A}}(V \otimes \mathbf{A})$$

defines a covariant functor from the category of superalgebras into the category of groups. We can compose this functor with the contravariant functor

$$M = (M_0, \mathcal{A}) \mapsto \mathcal{A}(M_0)$$

from the category of supermanifolds into that of superalgebras. The resulting contravariant functor

$$M \mapsto \mathrm{GL}(V)[M] := \mathrm{GL}_{\mathcal{A}(M_0)}(V \otimes \mathcal{A}(M_0))$$

from the category of supermanifolds into that of groups is denoted by $\mathrm{GL}(V)[\cdot]$.

DEFINITION 29. *A* **supergroup** $G[\cdot]$ *is a contravariant functor from the category of supermanifolds into that of groups. The supergroup* $\mathrm{GL}(V)[\cdot]$*, defined above, is called the* **general linear supergroup** *over the* $\mathbb{Z}_2$*-graded vector space* V*. A* **subgroup** *of a supergroup* $G[\cdot]$ *is a supergroup* $H[\cdot]$ *such that* $H[M]$ *is a subgroup of* $G[M]$ *for all supermanifolds* M *and* $H[\varphi] = G[\varphi]|H[N]$ *for all morphisms* $\varphi : M \to N$*. We will write* $H[\cdot] \subset G[\cdot]$ *if* $H[\cdot]$ *is a subgroup of the supergroup* $G[\cdot]$*. A* **linear supergroup** *is a subgroup* $H[\cdot] \subset \mathrm{GL}(V)[\cdot]$*.*

Example 7: To any linear super Lie algebra $\mathfrak{g} \subset \mathfrak{gl}(V)$ we can associate the linear supergroup $G[\cdot] \subset \mathrm{GL}(V)[\cdot]$ defined by:

$$G[M] := \langle \exp(\mathfrak{g} \otimes \mathcal{A}(M_0))_0 \rangle$$

for any supermanifold $M = (M_0, \mathcal{A})$. The right-hand side is the subgroup of $\mathrm{GL}_{\mathcal{A}(M_0)}(V \otimes \mathcal{A}(M_0))$ generated by the exponential image of the Lie algebra $(\mathfrak{g} \otimes \mathcal{A}(M_0))_0 \subset (\mathfrak{gl}(V) \otimes \mathcal{A}(M_0))_0 \cong (\mathrm{End}_{\mathcal{A}(M_0)}(V \otimes \mathcal{A}(M_0)))_0$. The convergence of the exponential series, which is locally uniform in all derivatives of arbitrary order, follows from the analyticity of the exponential map by a (finite) Taylor expansion with respect to the odd coordinates. $G[\cdot]$ is called the **linear supergroup associated to the linear super Lie algebra** $\mathfrak{g}$.

For any supermanifold M we denote by $\Delta_M = (\Delta_{M_0}, \Delta_M^*)$ the **diagonal embedding** defined by: $\Delta_{M_0}(x) = (x, x)$ $(x \in M_0)$ and $\Delta_M^* \pi_i^* f = f$ for all functions f on M. Here $\pi_i : M \times M \to M$ denotes the projection onto the i-th factor of $M \times M$ $(i = 1, 2)$.

DEFINITION 30. *A* **Lie supergroup** *is a supermanifold* $G = (G_0, \mathcal{A}_G)$ *whose underlying manifold is a Lie group* G_0 *(which, for convenience, we will always assume to be connected) with neutral element* $e \in G_0$*, multiplication* $\mu_0 : G_0 \times G_0 \to G_0$ *and inversion* $\iota_0 : G_0 \to G_0$*, together with morphisms* $\mu = (\mu_0, \mu^*) : G \times G \to G$ *and* $\iota = (\iota_0, \iota^*) : G \to G$ *such that*

i) $\mu \circ (\mathrm{Id}_G \times \mu) = \mu \circ (\mu \times \mathrm{Id}_G) \in \mathrm{Mor}(G \times G \times G, G)$,
ii) $\mu \circ (\mathrm{Id}_G \times \epsilon_e) = \pi_1 : G \times \{e\} \xrightarrow{\sim} G$, $\mu \circ (\epsilon_e \times \mathrm{Id}_G) = \pi_2 : \{e\} \times G \xrightarrow{\sim} G$ *and*
iii) $\mu \circ (\mathrm{Id}_G \times \iota) \circ \Delta_G = \mu \circ (\iota \times \mathrm{Id}_G) \circ \Delta_G = \mathrm{Id}_G$.

Here $\epsilon_e = (e, \epsilon_e^*) : \{e\} \hookrightarrow G_0 \overset{\epsilon}{\hookrightarrow} G$ *is the canonical embedding. The morphisms* μ *and* ι *are called, respectively,* **multiplication** *and* **inversion** *in the Lie supergroup* G*. A* **Lie subgroup** *of a Lie supergroup* $G = (G_0, \mathcal{A}_G)$ *is a submanifold* $\varphi : H = (H_0, \mathcal{A}_H) \hookrightarrow G$ *such that the immersion* $\varphi_0 : H_0 \hookrightarrow G_0$ *induces on* H_0 *the structure of Lie subgroup of* G_0 *and* φ *induces on* H *the structure of Lie supergroup with underlying Lie group* H_0*. We will write* $H \subset G$ *if* H *is a Lie subgroup of* G*. An* **action** *of a Lie supergroup* G *on a supermanifold* M *is a morphism* $\alpha : G \times M \to M$ *such that*

i) $\alpha \circ (\mathrm{Id}_G \times \alpha) = \alpha \circ (\mu \times \mathrm{Id}_M)$ *and*
ii) $\alpha \circ (\epsilon_e \times \mathrm{Id}_M) = \pi_2 : \{e\} \times M \xrightarrow{\sim} M$.

Given an action $\alpha : G \times M \to M$ we have the notion of **fundamental vector field** X associated to $x \in T_eG$. It is defined by $X(f) := x(\alpha^* f)$. The correspondence $x \mapsto X$ defines an even $\mathbb{R}$-linear map $T_eG \to \mathcal{T}_M(M_0)$ from the $\mathbb{Z}_2$-graded vector space T_eG to the free $\mathcal{A}_M(M_0)$-module of global vector fields on M.

DEFINITION 31. *Let an action α of a Lie supergroup G on a supermanifold M be given. A tensor field S on M is called* **G-invariant** *if the Lie derivative $\mathcal{L}_X S = 0$ for all fundamental vector fields X. (Notice that, in particular, this defines the notion of G-invariant pseudo-Riemannian metric on M.) An almost quaternionic structure Q on M is called* **G-invariant** *if $\mathcal{L}_X S$ is a section of Q for all sections S of Q and fundamental vector fields X on M.*

We can specialize the above definition to the (left-) action $\mu : G\times G \to G$ given by the multiplication in the Lie supergroup G. The G-invariant tensor fields with respect to that action are called **left-invariant**. A tensor field S is called **right-invariant** if $L_X S = 0$ for all left-invariant vector fields X on G. The right-invariant vector fields are precisely the fundamental vector fields for the action μ.

For any Lie supergroup G, we can define a group homomorphism $\mu_l : G_0 \to \mathrm{Aut}(G)$ by:

$$\mu_l(g) := \mu \circ (\epsilon_g \times \mathrm{Id}_G) : \{g\} \times G \cong G \to G\,, \quad g \in G_0\,.$$

Here $\epsilon_g : \{g\} \hookrightarrow G$ is the canonical embedding and $\{g\} \times G$ is canonically identified with G via the projection $\pi_2 : \{g\} \times G \xrightarrow{\sim} G$ onto the second factor. Similarly, we can define a group antihomomorphism $\mu_r : G_0 \to \mathrm{Aut}(G)$ by:

$$\mu_r(g) := \mu \circ (\mathrm{Id}_G \times \epsilon_g) : G \times \{g\} \cong G \to G\,, \quad g \in G_0\,.$$

Notice that the canonical embedding $\epsilon : G_0 \hookrightarrow G$ is G_0-equivariant with respect to the usual left- (respectively, right-) action on G_0 and the action on G defined by μ_l (respectively, μ_r).

Given a Lie supergroup G and a supermanifold M there is a natural group structure on $\mathrm{Mor}(M, G)$ with multiplication defined by:

$$\varphi \cdot \psi := \mu \circ (\varphi \times \psi) \circ \Delta_M\,, \quad \varphi, \psi \in \mathrm{Mor}(M, G)\,.$$

The correspondence $M \mapsto \mathrm{Mor}(M, G)$ defines a supergroup.

DEFINITION 32. *The supergroup $G[\cdot] := \mathrm{Mor}(\cdot, G)$ is called the supergroup* **subordinate** *to the Lie supergroup G.*

Example 8: Let V be a $\mathbb{Z}_2$-graded vector space. Then $\mathrm{GL}_{\mathbb{R}}(V)$ is, by definition, the group of invertible elements of $(\mathrm{End}_{\mathbb{R}} V)_0$. The Lie group $\mathrm{GL}_{\mathbb{R}}(V) \cong \mathrm{GL}_{\mathbb{R}}(V_0) \times \mathrm{GL}_{\mathbb{R}}(V_1)$ is an open submanifold of the vector space $(\mathrm{End}_{\mathbb{R}} V)_0 = SM(\mathrm{End}_{\mathbb{R}} V)_0$, see Example 5. We define the submanifold

$$\mathrm{GL}(V) := SM(\mathrm{End}_{\mathbb{R}} V)|_{\mathrm{GL}_{\mathbb{R}}(V)} \hookrightarrow SM(\mathrm{End}_{\mathbb{R}} V)\,.$$

The manifold underlying the supermanifold $\mathrm{GL}(V) = (\mathrm{GL}(V)_0, \mathcal{A}_{\mathrm{GL}(V)})$ is the above Lie group: $\mathrm{GL}(V)_0 = \mathrm{GL}_{\mathbb{R}}(V)$. From the definition of the supermanifold $\mathrm{GL}(V)$ it is clear that $\mathrm{Mor}(M, \mathrm{GL}(V))$ is canonically identified with the set $\mathrm{GL}_{\mathcal{A}(M_0)}(V \otimes \mathcal{A}(M_0)) = \mathrm{GL}(V)[M]$ (cf. Def. 29) for any supermanifold $M = (M_0, \mathcal{A})$. Moreover, $\mathrm{GL}(V)$ has a unique structure of Lie supergroup inducing the canonical group structure on $\mathrm{GL}(V)[M]$ for any supermanifold M. In other words, the general linear supergroup $\mathrm{GL}(V)[\cdot]$ is the supergroup subordinate to the Lie supergroup $\mathrm{GL}(V)$, see Def. 32.

Definition 33. *The Lie supergroup* $\mathrm{GL}(V)$ *is called the* **general linear Lie supergroup**. *A* **linear Lie supergroup** *is a Lie subgroup of* $\mathrm{GL}(V)$.

There exists a unique morphism

$$\mathrm{Exp} : SM(\mathfrak{gl}(V)) \to \mathrm{GL}(V)$$

such that

$$\mathrm{Exp} \circ \varphi = \exp(\varphi)$$

for all supermanifolds $M = (M_0, \mathcal{A})$ and all morphisms

$$\varphi \in \mathrm{Mor}(M, SM(\mathfrak{gl}(V))) = (\mathfrak{gl}(V) \otimes \mathcal{A}(M_0))_0 = (\mathrm{End}_{\mathcal{A}(M_0)} V \otimes \mathcal{A}(M_0))_0 ,$$

where, on the right-hand side,

$$\exp : (\mathrm{End}_{\mathcal{A}(M_0)} V \otimes \mathcal{A}(M_0))_0 \to \mathrm{GL}_{\mathcal{A}(M_0)}(V \otimes \mathcal{A}(M_0)) = \mathrm{Mor}(M, \mathrm{GL}(V))$$

is the exponential map for even endomorphisms of $V \otimes \mathcal{A}(M_0)$ (for the definition of the supermanifold $SM(\mathfrak{gl}(V))$ see Example 5). The underlying map Exp_0 : $\mathfrak{gl}(V)_0 = SM(\mathfrak{gl}(V))_0 \to \mathrm{GL}(V)_0$ is the ordinary exponential map for the Lie group $\mathrm{GL}(V)_0$: $\mathrm{Exp}_0 = \exp$. The morphism Exp is called the **exponential morphism** of $\mathfrak{gl}(V)$.

Example 9: Let $\mathfrak{g} \subset \mathfrak{gl}(V)$ be a linear super Lie algebra and $G[\cdot] \subset \mathrm{GL}(V)[\cdot]$ the correponding linear supergroup, see Example 7. We denote by $G_0 \subset GL(V)_0$ the connected linear Lie group with Lie algebra $\mathfrak{g}_0$. It is the immersed Lie subgroup generated by the exponential image of $\mathfrak{g}_0 = \mathfrak{gl}(V)_0$. We define an ideal $\mathcal{J}_g \subset (\mathcal{A}_{\mathrm{GL}(V)})_g$ as follows: A germ of function $f \in (\mathcal{A}_{\mathrm{GL}(V)})_g$ $(g \in G_0)$ belongs to $\mathcal{J}_g$ if and only if $r(\varphi^* f) = 0$ for all injective immersions $\varphi \in G[M] \subset Mor(M, \mathrm{GL}(V))$, where $r : ((\varphi_0)_* \mathcal{A}_M)_g \to (\mathcal{A}_M)_{\varphi_0^{-1}(g)}$ is the natural restriction map. We claim that $\mathcal{J}_G := \cup_{g \in G_0} \mathcal{J}_g$ is the vanishing ideal of a submanifold $G \hookrightarrow \mathrm{GL}(V)$. Due to the invariance of $\mathcal{J}_G$ under the group $\mu_l(G_0) \subset \mathrm{Aut}(G)$ it is sufficient to prove the claim locally over an open set of the form $\exp U \subset G_0$, $U \subset \mathfrak{g}_0$ an open neighborhood of $0 \in \mathfrak{g}_0$. The local statement follows from the fact that the morphism $SM(\mathfrak{g}) \hookrightarrow SM(\mathfrak{gl}(V)) \stackrel{\mathrm{Exp}}{\to} \mathrm{GL}(V)$ has maximal rank at $0 \in \mathfrak{g}_0 = SM(\mathfrak{g})_0$ and hence defines a submanifold $SM(\mathfrak{g})|_U \hookrightarrow \mathrm{GL}(V)$ for some open neighborhood of U of $0 \in \mathfrak{g}_0$. The vanishing ideal of this submanifold coincides with $\mathcal{J}_G$ over $\exp U$ by the definition of the exponential morphism. Next we claim that multiplication $\mu : \mathrm{GL}(V) \times \mathrm{GL}(V) \to \mathrm{GL}(V)$ and inversion $\iota : \mathrm{GL}(V) \to \mathrm{GL}(V)$ have the property that $\mu^* \mathcal{J}_G \subset \mathcal{J}_{G \times G}$ and $\iota^* \mathcal{J}_G = \mathcal{J}_G$. Here $\mathcal{J}_{G \times G}$ is the vanishing ideal of the submanifold $G \times G \hookrightarrow \mathrm{GL}(V) \times \mathrm{GL}(V)$. The claim follows from the fact that $G[\cdot]$ is a subgroup of $\mathrm{GL}(V)[\cdot]$ and implies that the morphisms $G \times G \hookrightarrow \mathrm{GL}(V) \times \mathrm{GL}(V) \stackrel{\mu}{\to} \mathrm{GL}(V)$ and $G \hookrightarrow \mathrm{GL}(V) \stackrel{\iota}{\to} \mathrm{GL}(V)$ induce morphisms $G \times G \to G$ and $G \to G$, which induce on G the structure of Lie supergroup. In other words, G is a Lie subgroup of the general linear Lie supergroup $\mathrm{GL}(V)$. It is called the **linear Lie supergroup associated to the linear Lie superalgebra** $\mathfrak{g} \subset \mathfrak{gl}(V)$. Notice that Exp^* maps $\mathcal{J}_G$ into the vanishing ideal of $SM(\mathfrak{g}) \hookrightarrow SM(\mathfrak{gl}(V))$ and hence the restriction $SM(\mathfrak{g}) \hookrightarrow SM(\mathfrak{gl}(V)) \stackrel{\mathrm{Exp}}{\to} \mathrm{GL}(V)$ of the exponential morphism induces a morphism $SM(\mathfrak{g}) \to G$, which is called the **exponential morphism** of $\mathfrak{g}$ and is again denoted by Exp. Its differential at $0 \in \mathfrak{g}_0$ yields an isomorphism $\mathfrak{g} \cong T_0 SM(\mathfrak{g}) \stackrel{\sim}{\to} T_e G$.

4.7. Homogeneous supermanifolds. Let $G = (G_0, \mathcal{A}_G)$ be a Lie supergroup, $K \subset G_0$ a closed subgroup and $\pi_0 : G_0 \to G_0/K$ the canonical projection. Then the subsheaf $\mathcal{A}_G^K := \mathcal{A}_G^{\mu_r(K)} \subset \mathcal{A}_G$ of functions on G invariant under the subgroup $\mu_r(K) \subset \mathrm{Aut}(G)$ is again a sheaf of superalgebras on G_0. Explicitly, a function $f \in \mathcal{A}_G(U)$ ($U \subset G_0$ open) belongs to $\mathcal{A}_G^K(U)$ if it can be extended to a $\mu_r(K)$-invariant function over $UK \subset G_0$. Its pushed forward sheaf $\mathcal{A}_{G/K} := (\pi_0)_* \mathcal{A}_G^K$ is a sheaf of superalgebras on the homogeneous manifold G_0/K.

THEOREM 18. *(cf. [**K**]) Let $G \hookrightarrow \mathrm{GL}(V)$ be a closed linear Lie supergroup and $K \subset G_0$ a closed subgroup. Then $G/K = (G_0/K, \mathcal{A}_{G/K})$ is a supermanifold with a canonical submersion $\pi : G \to G/K$ and a canonical action $\alpha : G \times G/K \to G/K$.*

Proof: Since $\mu_l(g) \in \mathrm{Aut}(G)$ induces isomorphisms

$$\mathcal{A}_{G/K}(gU) \xrightarrow{\sim} \mathcal{A}_{G/K}(U)$$

for all $g \in G_0$ and $U \subset G_0/K$ open, it is sufficient to check that $G/K|_U = (U, \mathcal{A}_{G/K}|_U)$ is a supermanifold for some open neighborhood U of $eK \in G_0/K$.

LEMMA 12. *Under the assumptions of Thm. 18, there exists local coordinates (x, y, ξ) for $\mathrm{GL}(V)$ over some neighborhood $U = UK \subset \mathrm{GL}(V)_0$ of $e \in \mathrm{GL}(V)_0$ such that*

1) *$x = (x^i)$ and $y = (y^j)$ consist of even functions and $\xi = (\xi^k)$ of odd functions,*
2) *$\mathcal{A}^K_{\mathrm{GL}(V)}(U)$ is the subalgebra of $\mathcal{A}_{\mathrm{GL}(V)}(U)$ which consists of functions $f(x,\xi)$ only of (x, ξ), more precisely,*

$$f(x,\xi) = \sum_\alpha f_\alpha(x)\xi^\alpha ,$$

where the $f_\alpha(x) \in \mathcal{A}_{\mathrm{GL}(V)}(U)$ are functions only of x, i.e. if $f_\alpha(x) \neq 0$ then $f_\alpha(x)$ does not belong to the ideal generated by ξ and $\epsilon^* f_\alpha(x) \in C^\infty(U)$ are functions only of $\epsilon^* x$ (independent of $\epsilon^* y$). Here we used the multiindex notation $\alpha = (\alpha_1, \dots, \alpha_{2mn}) \in \mathbb{Z}_2^{2mn}$ ($m|n = \dim V$) and $\xi^\alpha = \prod_{j=1}^{2mn} (\xi^j)^{\alpha_j}$.

Proof: The natural (global) coordinates on the supermanifold $\mathrm{GL}(V)$ are the matrix coefficients with respect to some basis of V. We will denote them simply by (z^i, ζ^j) instead of using matrix notation. For these coordinates it is clear that $\mu_r(g)^* z^k$ is a linear combination (over the real numbers) of the even coordinates $z := (z^i)$ and $\mu_r(g)^* \zeta^l$ is a linear combination (over the real numbers) of the odd coordinates $\zeta := (\zeta^j)$ for all $g \in \mathrm{GL}(V)_0$. In particular, we obtain a representation ρ of $K \subset \mathrm{GL}(V)_0$ on the vector space spanned by the odd coordinates. Let $E_\rho \to G_0/K$ denote the vector bundle associated to this representation. Any $\mu_r(K)$-invariant function on $\mathrm{GL}(V)$ linear in ζ defines a section of the dual vector bundle E_ρ^* and vice versa. Now, since E_ρ^* is locally trivial (like any vector bundle), we can find $\mu_r(K)$-invariant local functions $\xi = (\xi^j)$ linear in ζ such that (z, ξ) are local coordinates for $\mathrm{GL}(V)$ over some open neighborhood $U = UK \subset \mathrm{GL}(V)_0$ of e. Next, by way of a local diffeomorphism in the even coordinates z, we can arrange that $z = (x, y)$, where the x are $\mu_r(K)$-invariant functions on $\mathrm{GL}(V)$ such that $\epsilon^* x \in C^\infty(U)$ induce local coordinates on G_0/K. Now any function $f \in \mathcal{A}_{\mathrm{GL}(V)}(U)$ has a unique expression of the form

$$\sum_\alpha f_\alpha(x, y)\xi^\alpha , \tag{23}$$

where the $f_\alpha(x,y)$ are functions only of (x,y). From the $\mu_r(K)$-invariance of the ξ^α it follows that f is $\mu_r(K)$-invariant if and only if the $f_\alpha(x,y)$ are $\mu_r(K)$-invariant. The function $f_\alpha(x,y)$ is $\mu_r(K)$-invariant if and only if $\epsilon^* f_\alpha(x,y) \in C^\infty(U)$ is invariant under the right-action of K on GL_0, i.e. if and only if $\epsilon^* f_\alpha(x,y)$ is a function only of $\epsilon^* x$. This shows that $f \in \mathcal{A}_{\mathrm{GL}(V)}(U)^K$ if and only if the coefficients $f_\alpha(x,y)$ in the expansion (23) are functions only of x $\square$

We continue the proof of Thm. 18. From Lemma 12 it follows that $\mathrm{GL}(V)/K$ is a supermanifold. In fact, the K-invariant local functions (x,ξ) on $\mathrm{GL}(V)$ constructed in that lemma induce local coordinates on $\mathrm{GL}(V)/K$. Next we restrict the coordinates (x,y,ξ) to the submanifold $G \hookrightarrow \mathrm{GL}(V)$. We can decompose $x = (x', x'')$, $\xi = (\xi', \xi'')$ such that (x', y, ξ') restrict to local coordinates on G over some open neighborhood (again denoted by) U of $e \in G_0$ (notice that, by construction, y restrict to local coordinates on K). Now, as in the proof of the corresponding statement for $\mathrm{GL}(V)$ (see Lemma 12), it follows that $\mathcal{A}_G^K(U)$ consists precisely of all functions of the form $f(x',\xi') = \sum f_\alpha(x')(\xi')^\alpha$. This proves that G/K is a supermanifold with local coordinates (x',ξ') over U.

The inclusion $\mathcal{A}_G^K \subset \mathcal{A}_G$ defines the canonical submersion $\pi : G \to G/K$. Finally, if f is a $\mu_r(K)$-invariant (local) function on G then $\mu^* f$ is a (local) function on $G\times G$ invariant under the group $\mathrm{Id}_G \times \mu_r(K) \subset \mathrm{Aut}(G\times G)$. This shows that the composition $G \times G \xrightarrow{\mu} G \xrightarrow{\pi} G/K$ factorizes to a morphism $\alpha : G \times G/K \to G/K$, which defines an action of G on G/K. $\square$

DEFINITION 34. *The supermanifold $M = G/K$ is called the* **homogeneous supermanifold** *associated to the pair (G,K).*

For the rest of the paper let $\mathfrak{g} \subset \mathfrak{gl}(V)$ be a linear super Lie algebra, $\mathfrak{k} \subset \mathfrak{g}_0$ a subalgebra and $\mathfrak{g} = \mathfrak{k} + \mathfrak{m}$ a $\mathfrak{k}$-invariant direct decomposition compatible with the $\mathbb{Z}_2$-grading. We denote by $K \subset G_0 \subset G \subset \mathrm{GL}(V)$ the corresponding linear Lie supergroups (see Example 9) and assume that the (connected) Lie subgroups $K \subset G_0 \subset \mathrm{GL}(V)_0$ are closed. Then, by Thm. 18, $M = G/K$ is a supermanifold with a canonical action of G. We have the canonical identification $\mathfrak{m} \cong \mathfrak{g}/\mathfrak{k} \cong T_{eK}M$ given by $x \mapsto X(eK)$, where $X(eK)$ is the value of the fundamental vector field X on M associated to $x \in \mathfrak{m}$ at the base point $eK \in G_0/K = M_0$. We claim that any $\mathrm{ad}_\mathfrak{k}$-invariant tensor S_{eK} over $\mathfrak{m}$ defines a correponding G-invariant (see Def. 31) tensor field on M such that $S(eK) = S_{eK}$. Here by a tensor over $\mathfrak{m}$ we mean an element of the full tensor superalgebra $\otimes\langle \mathfrak{m}, \mathfrak{m}^* \rangle$ over $\mathfrak{m}$. In fact, for any tensor S_{eK} over $\mathfrak{m}$ there exists a corresponding left-invariant tensor field S on G such that $S(eK) = S_{eK}$. In order for S to define a tensor field on G/K it is necessary and sufficient that S is $\mu_r(K)$-invariant, or, equivalently, that S_{eK} is $\mathrm{ad}_\mathfrak{k}$-invariant. In particular, we have the following proposition:

PROPOSITION 16. *Let g_{eK} be an $\mathrm{ad}_\mathfrak{k}$-invariant nondegenerate supersymmetric bilinear form on $\mathfrak{m}$. Then there exists a unique G-invariant pseudo-Riemannian metric g on $M = G/K$ (see Def. 23 and Def. 31) such that $g(eK) = g_{eK}$. Let Q_{eK} be an $\mathrm{ad}_\mathfrak{k}$-invariant quaternionic structure on $\mathfrak{m}$ (i.e. $\mathrm{ad} : \mathfrak{k} \to \mathfrak{gl}(\mathfrak{m})$ normalizes Q_{eK}). Then there exists a unique G-invariant almost quaternionic structure Q on M (see Def. 24 and Def. 31) such that $Q(eK) = Q_{eK}$.*

Finally, we need to discuss G-invariant connections on $M = G/K$.

DEFINITION 35. *A connection ∇ on a homogeneous supermanifold $M = G/K$ is called* G-**invariant** *if*

$$\mathcal{L}_X(\nabla_Y S) = \nabla_{[X,Y]} S + (-1)^{\tilde{X}\tilde{Y}} \nabla_Y(\mathcal{L}_X S)$$

for all vector fields X and Y on M.

Let ∇ be a connection on a supermanifold M. For any vector field X on M one defines the $\mathcal{A}_M(M_0)$-linear operator

$$L_X := \mathcal{L}_X - \nabla_X \,.$$

We denote by $L_X(p) \in \mathrm{End}_{\mathbb{R}}\, T_pM$ its value at $p \in M_0$; it is defined by $L_X(p)Y(p) = (L_XY)(p)$ for all vector fields Y on M.

For a G-invariant connection ∇ on a homogeneous supermanifold $M = G/K$ as above we define the **Nomizu map** $L = L(\nabla) : \mathfrak{g} \to \mathrm{End}(T_{eK}M)$, $x \mapsto L_x$, by the equation

$$L_x := L_X(eK)\,,$$

where X is the fundamental vector field on M associated to $x \in \mathfrak{g}$. The operators $L_x \in \mathrm{End}(T_{eK}M)$ will be called **Nomizu operators**. They have the following properties:

$$L_x = d\rho(x) \quad \text{for all} \quad x \in \mathfrak{k} \tag{24}$$

and

$$L_{\mathrm{Ad}_k x} = \rho(k) L_x \rho(k)^{-1} \quad \text{for all} \quad x \in \mathfrak{g}\,, \quad k \in K\,, \tag{25}$$

where $\rho : K \to \mathrm{GL}(T_{eK}M)$ is the isotropy representation (under the identification $T_{eK}M \cong \mathfrak{m}$ the representation ρ is identified with adjoint representation of K on $\mathfrak{m}$).

Conversely, any even linear map $L : \mathfrak{g} \to \mathrm{End}(T_{[e]}M)$ satisfying (24) and (25) is the Nomizu map of a uniquely defined G-invariant connection $\nabla = \nabla(L)$ on M. Its torsion tensor T and curvature tensor R are expressed at eK by:

$$T(\pi x, \pi y) = -(L_x \pi y - (-1)^{\tilde{x}\tilde{y}} L_y \pi x + \pi[x,y])$$

and

$$R(\pi x, \pi y) = [L_x, L_y] + L_{[x,y]}\,, \quad x, y \in \mathfrak{g}$$

where $\pi : \mathfrak{g} \to T_{eK}M$ is the canonical projection

$$x \mapsto \pi x = X(eK) = \frac{d}{dt}|_{t=0} \exp(tx)K\,.$$

Suppose now that we are given a G-invariant geometric structure S on M (e.g. a G-invariant almost quaternionic structure Q) defined by a corresponding K-invariant geometric structure S_{eK} on $T_{eK}M$. Then a G-invariant connection ∇ preserves S if and only if the corresponding Nomizu operators L_x, $x \in \mathfrak{g}$, preserve S_{eK}. So to construct a G-invariant connection preserving S it is sufficient to find a Nomizu map $L : \mathfrak{g} \to \mathrm{End}(T_{eK}M)$ such that L_x preserves S_e for all $x \in \mathfrak{g}$. We observe that, due to the K-invariance of S_e, the Nomizu operators L_x preserve S_e already for $x \in \mathfrak{k}$. The above considerations can be specialized as follows:

PROPOSITION 17. *Let Q be a G-invariant almost quaternionic structure on a homogeneous supermanifold $M = G/K$. There is a natural one-to-one correspondence between G-invariant almost quaternionic connections on (M, Q) and Nomizu maps $L : \mathfrak{g} \to \mathrm{End}(T_{eK}M)$, whose image normalizes $Q(eK)$, i.e. whose Nomizu operators L_x, $x \in \mathfrak{g}$, belong to the normalizer $\mathrm{n}(Q) \cong \mathfrak{sp}(1) \oplus \mathfrak{gl}(d, \mathbb{H})$ $(d = (m+n)/4)$ of the quaternionic structure $Q(eK)$ in the super Lie algebra $\mathfrak{gl}(T_{eK}M)$.*

COROLLARY 10. *Let $(M = G/K, Q)$ be a homogeneous almost quaternionic supermanifold and $L : \mathfrak{g} \to \mathrm{End}(T_{eK}M)$ a Nomizu map such that*

(1) *$L_x \pi y - (-1)^{\tilde{x}\tilde{y}} L_y \pi x = -\pi[x, y]$ for all $x, y \in \mathfrak{g}$ (i.e. $T = 0$) and*
(2) *L_x normalizes $Q(eK) \subset \mathrm{End}(T_{eK}M)$.*

Then $\nabla(L)$ is a G-invariant quaternionic connection on (M, Q) and hence Q is 1-integrable.

For use in 4, we give the formula for the Nomizu map L^g associated to the Levi-Civita connection ∇^g of a G-invariant pseudo-Riemannian metric g on a homogeneous supermanifold $M = G/K$. Let $\langle \cdot, \cdot \rangle = g(eK)$ be the K-invariant nondegenerate supersymmetric bilinear form on $T_{eK}M$ induced by g (the value of g at eK). Then $L^g_x \in \mathrm{End}(T_{eK}M)$, $x \in \mathfrak{g}$, is given by the following Koszul type formula:

(26)
$$-2\langle L^g_x \pi y, \pi z\rangle = \langle \pi[x, y], \pi z\rangle - \langle \pi x, \pi[y, z]\rangle - (-1)^{\tilde{x}\tilde{y}} \langle \pi y, \pi[x, z]\rangle \,, \quad x, y, z \in \mathfrak{g} \,.$$

COROLLARY 11. *Let $(M = G/K, Q, g)$ be a homogeneous almost quaternionic (pseudo-) Hermitian supermanifold and assume that L^g_x normalizes $Q(eK)$ for all $x \in \mathfrak{g}$. Then the Levi-Civita connection $\nabla^g = \nabla(L^g)$ is a G invariant quaternionic connection on (M, Q, g) and hence (M, Q, g) is a quaternionic (pseudo-) Kähler supermanifold if* $\dim M_0 > 4$.

Bibliography

[A1] D.V. Alekseevskiĭ: *Riemannian spaces with exceptional holonomy groups*, Funct. Anal. Applic. **2** (1968), 97-105

[A2] D.V. Alekseevskiĭ: *Compact quaternion spaces*, Funct. Anal. Applic. **2** (1968), 106-114

[A3] D.V. Alekseevskiĭ: *Classification of quaternionic spaces with a transitive solvable group of motions*, Math. USSR Izvestija **9**, No. 2 (1975), 297–339

[A4] D.V. Alekseevskiĭ: *Conjugacy of polar factorizations of Lie groups*, Math. USSR. Sbornik **13**, No. 1 (1971), 12–25

[A5] D.V. Alekseevsky: *Flag manifolds*, in "11th Yugoslav Geometrical Seminar" (Divcibare, 1996), Zb. Rad. Mat. Inst. Beograd. (N.S.) **6** (14) (1997), 3-35

[A-C1] D.V. Alekseevsky, V. Cortés: *Isometry groups of homogeneous quaternionic Kähler manifolds*, Preprint Erwin Schrödinger Institut 230 (1995) (44 pp.), to appear in Journal of Geometric Analysis

[A-C2] D.V. Alekseevsky, V. Cortés: *Classification of N-(super)-extended Poincaré algebras and bilinear invariants of the spinor representation of $Spin(p,q)$*, Commun. Math. Phys. **183** (1997), 477-510

[A-C3] D.V. Alekseevsky, V. Cortés: *Mirror supersymmetry in the space of (super) extended Poincaré algebras*, Lett. Math. Phys. **38** (1996), no. 3, 283-287

[A-C4] D.V. Alekseevsky, V. Cortés: *Homogeneous quaternionic Kähler manifolds of unimodular group*, Boll. Un. Mat. Ital. B (7) **11** (1997), no. 2, suppl. 217-229

[A-C5] D.V. Alekseevsky, V. Cortés: *Classification of stationary compact homogeneous special pseudo Kähler manifolds of semisimple group*, SFB 256 "Nichtlineare Partielle Differentialgleichungen", Bonn, Preprint 519 (23 pp.), to appear in Proc. London Math. Soc.

[A-C-D-S] D.V. Alekseevsky, V. Cortés, C. Devchand, U. Semmelmann: *Killing spinors are Killing vector fields in Riemannian supergeometry*, J. Geom. Phys. **26** (1998), 37-50

[A-G] D.V. Alekseevskiĭ, M.M. Graev: *Grassmann and hyper-Kähler structures on some spaces of sections of holomorphic bundles*, in "Manifolds and geometry" (Pisa, 1993), 1-19, Sympos. Math., XXXVI, Cambridge Univ. Press, Cambridge, 1996

[A-K] D.V. Alekseevskiĭ, B.N. Kimel'fel'd: *Structure of homogeneous Riemannian spaces with zero Ricci curvature*, Funct. Anal. Applic. **9** (1975), 97-102

[A-M1] D.V. Alekseevsky, S. Marchiafava: *Transformations of a quaternionic Kähler manifold*, C. R. Acad. Sci. Paris Ser. I Math. **320** (1995), no. 6, 703-708

[A-M2] D.V. Alekseevsky, S. Marchiafava: *Quaternionic structures on a manifold and subordinate structures*, Ann. Mat. Pura Appl. (4) **171** (1996), 205-273

[A-M3] D.V. Alekseevsky, S. Marchiafava: *Quaternionic transformations of a non-positive quaternionic Kähler manifold*, Internat. J. Math. **8** (1997), no. 3, 301-316

[A-M-P] D.V. Alekseevsky, S. Marchiafava, M. Pontecorvo: *Compatible complex structures on almost quaternionic manifolds*, Preprint Erwin Schrödinger Institut 404 (1996) (19 pp.), to appear in Tans. Amer. Math. Soc.

[A-P] D.V. Alekseevsky, F. Podestà: *Compact cohomogeneity one Riemannian manifolds of positive Euler characteristic and quaternionic Kähler manifolds*, Preprint Erwin Schrödinger Institut 466 (1997) (32 pp.)

[A-V-L] D.V. Alekseevsky, A.M. Vinogradov, V.V. Lychagin: *Geometry I*, EMS **28**, Springer, Berlin, Heidelberg, New York, 1991

[A-H-S] M.F. Atiyah, N.J. Hitchin, I.M. Singer: *Self-duality in four-dimensional Riemannian geometry*, Proc. Roy. Soc. London A **362** (1978), 425-461

[A-W] R. Azencott, E. Wilson: *Homogeneous manifolds with negative curvature II*, Mem. Am. Math. Soc. **178** (1976)

[B-W] J. Bagger, E. Witten: *Matter couplings in $N = 2$ supergravity*, Nucl. Phys. **B222** (1983), 1-10

[B-D] M.L. Barberis, I. Dotti Miatello: *Hypercomplex structures on a class of solvable Lie groups*, Quart. J. Math. Oxford Ser. (2) **47** (1996), no. 188, 389-404

[B-B-H] C. Bartocci, U. Bruzzo, D. Hernández-Ruipérez: *The geometry of supermanifolds*, Kluwer Academic Publishers Group, Dordrecht, 1991

[Ba] M. Batchelor: *The structure of supermanifolds*, Trans. Amer. Math. Soc. **253** (1979), 329-338

[Bea] A. Beauville: *Fano contact manifolds and nilpotent orbits*, preprint alg-geom/9707015

[Be] F.A. Berezin: *Introduction to superanalysis*, MPAM, D. Reidel, Dordrecht, 1987

[Bern] J. Bernstein: *Lectures on supersymmetry*, notes by P. Deligne and J. Morgan

[Bes] A. Besse: *Einstein manifolds*, Ergebnisse der Mathematik und ihrer Grenzgebiete, Springer Verlag, Berlin, 1984

[Bi1] R. Bielawski: *Invariant hyperKähler metrics with a homogeneous complex structure*, Math. Proc. Camb. Phil. Soc. **122** (1997), 473-482

[Bi2] R. Bielawski: *Complete hyperKähler $4n$-manifolds with n commuting tri-Hamiltonian vector fields*, preprint http://xxx.lanl.gov/abs/math/9808134

[B-G] O. Biquard, P. Gauduchon: *La métrique hyperkählerienne des Orbites coadjointes de type symétrique d'un group de Lie complexe semi-simple*, C. R. Acad. Sci. Paris Ser. I Math. **323** (1996), no. 12, 1259-1264

[Bo] A. Borel: *Compact Clifford-Klein forms of symmetric spaces*, Topology **2** (1963), 111-122

[Br1] R. Bryant: *A survey of Riemannian metrics with special holonomy groups*, Proc. ICM Berkeley, 1986, Amer. Math. Soc., 1987, 505-514

[Br2] R. Bryant: *Classical, exceptional and exotic holonomies: a status report*, Actes de la Table Ronde de Géométrie Différentielle en l'Honneur de Marcel Berger, Collection SMF Séminairs et Congrès **1** (1996), 93-166

[B-O] V.A. Bunegina, A.L. Onishchik: *Two families of flag supermanifolds*, Diff. Geom. Appl. **4** (1994), no. 4, 329-360

[C-D-F] L. Castellani, R. D'Auria, S. Ferrara: *Special geometry without special coordinates*, Class. Quantum Grav. **7** (1990), 1767-1790

[Ce] S. Cecotti: *Homogeneous Kähler manifolds and T-algebras in $N = 2$ supergravity and superstrings*, Commun. Math. Phys. **124** (1989), 23-55

[C-F-G] S. Cecotti, S. Ferrara, L. Girardello: *Geometry of type II superstrings and the moduli of superconformal field theories*, Intern. J. Mod. Phys. **A4** (1989), 2475-2529

[C1] V. Cortés: *Alekseevskiĭs quaternionische Kählermannigfaltigkeiten*, Dissertation, Bonner Mathematische Schriften **267** (1994)

[C2] V. Cortés: *Alekseevskian spaces*, Diff. Geom. Appl. **6** (1996), 129-168

[C3] V. Cortés: *Homogeneous special geometry*, Transform. Groups **1** (1996), no. 4, 337-373

[C4] V. Cortés: *On hyper Kähler manifolds associated to Lagrangian Kähler submanifolds of $T^*\mathbb{C}^n$*, Trans. Amer. Math. Soc. **350** (1998), no. 8, 3193-3205

[C5] V. Cortés: *What is the role of twistors in supergeometry?*, Proceedings of the Third International Workshop on Differential Geometry and its Applications and the First German-Romanian Seminar on Geometry, Sibiu/Hermannstadt, 1997; General Math. **5** (1997), 125-130

[Cr] E. Cremmer: *Dimensional reduction in field theory and hidden symmetries in extended supergravity*, in Supergravity '81 (eds. S. Ferrara, J.G. Taylor), Cambridge University Press, 1982, 313-368

[C-VP] E. Cremmer, A. Van Poeyen: *Classification of Kähler manifolds in $N = 2$ vector multiplet-supergravity couplings*, Class. Quantum Grav. **2** (1985), 445-454

[D-D] J.E. D'Atri, I. Dotti Miatello: *A characterization of bounded symmetric domains by curvature*, Trans. Amer. Math. Soc. **276** (2) (1983), 531-540

[D-L] C. Devchand, O. Lechtenfeld: *Extended self-dual Yang-Mills from the $N = 2$ string*, preprint hep-th/9712043

[D-N] J. Dorfmeister, K. Nakajima: *The fundamental conjecture for homogeneous Kähler manifolds*, Acta Math. **161** (1988), no. 1-2, 23-70

[D-S1] A.S. Dancer, A.F. Swann: *Hyper-Kähler metrics associated to compact Lie groups*, Math. Proc. Cambridge Philos. Soc. **120** (1996), no. 1, 61-69

[D-S2] A.S. Dancer, A.F. Swann: *HyperKähler metrics of cohomogeneity one*, J. Geom. Phys. **21** (1997), 218-230

[D-S3] A.S. Dancer, A.F. Swann: *Quaternionic Kähler manifolds of cohomogeneity one*, preprint http://xxx.lanl.gov/abs/math/9808097

[DW] B. DeWitt: *Supermanifolds*, Cambridge University Press, Cambridge, 1992

[dW-VP1] B. de Wit, A. Van Proeyen: *Potentials and symmetries of general gauged $N = 2$ supergravity-Yang-Mills models*, Nucl. Phys. **B245** (1984), 89-117 [dW-VP2] B. de Wit, A. Van Proeyen: *Special geometry, cubic polynomials and homogeneous quaternionic spaces*, Commun. Math. Phys. **149** (1992), 307-333

[dW-V-VP] B. de Wit, F. Vanderseypen, A. Van Proeyen: *Symmetry structure of special geometries*, Nucl. Phys. **B400** (1993), 463-521

[F-S] S. Ferrara, S. Sabharwal: *Quaternionic manifolds for type II superstring vacua of Calabi-Yau spaces*, Nucl. Phys. **B332** (1990), 317-332

[Fi-S] A. Fino, S. Salamon: *Observations on the topology of symmetric spaces*, in "Geometry and Physics" (Aarhus, 1995), 275-286, Lecture Notes in Pure and Appl. Math. **184**, Dekker, New York, 1997

[Fr] D.S. Freed: *Special Kähler manifolds*, preprint hep-th/9712042

[F] P.G.O. Freund: *Introduction to supersymmetry*, Cambridge University Press, New York, 1986

[G1] K. Galicki: *Generalisation of the momentum mapping construction for quaternionic Kähler manifolds*, Commun. Math. Phys. **108** (1987), 117-138

[G2] K. Galicki: *Geometry of the scalar couplings in $N = 2$ supergravity models*, Class. Quantum Grav. **9** (1992), 27-40

[G-L] K. Galicki, H.B. Lawson: *Quaternionic reduction and quaternionic orbifolds*, Math. Ann. **282** (1988), 1-21

[G-S] K. Galicki, S. Salamon: *Betti numbers of 3-Sasakian manifolds*, Geom. Dedicata **63** (1996), no. 1, 45-68

[Gau] P. Gauduchon: *Canonical connections for almost-hypercomplex structures*, in "Complex analysis and geometry" (Trento, 1995), 123-136, Pitman Res. Notes Math. Ser. 366, Longman, Harlow, 1997

[G-M-P] G. Gentili, S. Marchiafava, M. Pontecorvo: Proceedings of the Meeting on Quaternionic Structures in Mathematics and Physics, Trieste, 1994; SISSA, Trieste, ILAS/FM-6, 1996, http://www.emis.de/proceedings/QSMP94/index.html

[G-P-V] S.G. Gindikin, I.I. Piatetskii-Shapiro, E.B. Vinberg: *Homogeneous Kähler manifolds*, in "Geometry of homogeneous bounded domains" (C.I.M.E., 3° Ciclo, Urbino, 1967), Edizioni Cremones, Roma, 1968

[Go-L] T.A. Gol'fand, E.P. Likhtman: *Extension of the algebra of Poincaré group generators and violation of P-invariance*, JETP Lett. **13** (1971), 323-326

[G-W] C.S. Gordon, E.N. Wilson: *Isometry groups of Riemannian solvmanifolds*, Trans. Amer. Math. Soc. **307** (1988), no. 1, 245-269

[G-S-T] M. Günaydin, G. Sierra, P.K. Townsend: *The geometry of $N = 2$ Maxwell-Einstein supergravity and Jordan algebras*, Nucl. Phys. **B242** (1984), 244-268

[H] S. Helgason: *Differential geometry, Lie groups and symmetric spaces*, Academic Press, Orlando, 1978

[H-K-L-R] N.J. Hitchin, A. Karlhede, U. Lindström, M. Roček: *Hyperkähler metrics and supersymmetry*, Commun. Math. Phys. **108** (1987), 535-589

[I-K] S. Ishihara, M. Konishi: *Fibered Riemannian spaces with Sasakian 3-structure*, Differential Geometry, in honor of K. Yano, Kinokuniya, Tokyo (1972), 179-194

[J1] D. Joyce: *The hypercomplex quotient and the quaternionic quotient*, Math. Ann. **290** (1991), no. 2, 323-340

[J2] D. Joyce: *Compact hypercomplex and quaternionic manifolds*, J. Differential Geom. **35** (1992), 743-761

[Kac] V. Kac: *Lie Superalgebras*, Adv. Math. **26** (1977), 8-96

[Kak] T. Kakolewski: *Der Twistorraum quaternionischer kählerscher Mannigfaltigkeiten*, Dissertation, Bonner Mathematische Schriften **280** (1995)

[K-R] W. Kühnel, H-B. Rademacher: *Asymptotically Euclidean manifolds and twistor spinors*, Max-Panck-Institut für Mathematik in den Naturwissenschaften, Leipzig, Preprint-Nr. 26 (1997)

[K-S1] P.Z. Kobak and A. Swann: *Quaternionic geometry of a nilpotent variety*, Math. Ann. **297** (1993), no. 4, 747-764

[K-S2] P.Z. Kobak and A. Swann: *Classical nilpotent orbits as hyper-Kähler quotients*, Internat. J. Math. **7** (1997), no. 2, 193-210

[K-NII] S. Kobayashi, K. Nomizu: *Foundations of differential geometry*, Vol. II, Interscience, New York, 1969

[Ko] M. Konishi: *On manifolds with Sasakian 3-structure over quaternionic Kählerian manifolds* Kodai Math. Sem. Reps. **26** (1975), 194-200

[K] B. Kostant: *Graded manifolds, graded Lie theory and prequantization* in Springer LNM **570** (1977), 177–306

[L-M] H.B. Lawson, M.-L. Michelson: *Spin geometry*, Princeton Univ. Press, 1989

[L1] C. LeBrun: *A rigidity theorem for quaternionic-Kähler manifolds*, Proc. Amer. Math. Soc. **103** (1988), no. 4, 1205-1208

[L2] C. LeBrun: *Quaternionic-Kähler manifolds and conformal geometry*, Math. Ann. **284** (1989), no. 3, 353-376

[L3] C. LeBrun: *On complete quaternionic-Kähler manifolds*, Duke Math. J. **63** (1991), no. 3, 723-743

[L4] C. LeBrun: *A finiteness theorem for quaternionic-Kähler manifolds with positive scalar curvature*, in "The Penrose transform and analytic cohomology in representation theory" (South Hadley, MA, 1992), 89-101, Contemp. Math. **154**, Amer. Math. Soc., Providence, 1993

[L5] C. LeBrun: *Fano manifolds, contact structures, and quaternionic geometry*, Internat. J. Math. **6** (1995), no. 3, 419-437

[L-S] C. LeBrun, S. Salamon: *Strong rigidity of positive quaternion-Kähler manifolds*, Invent. Math. **118** (1994), no. 1, 109-132

[L] D.A. Leites: *Introduction to the theory of supermanifolds*, Russian Math. Surveys **35** (1980), 3-57

[M] Yu.I. Manin: *Gauge field theory and complex geometry*, Springer-Verlag, Berlin-Heidelberg, 1988

[N-N] A. Newlander, L. Nirenberg: *Complex analytic coordinates in almost complex manifolds*, Ann. of Math. **65** (1957), 391-404

[O-S] V.I. Ogievetsky, E.S. Sokatchev: *Structure of supergravity group*, Phys. Lett. **79B** (1978), 222-224

[O'N] B. O'Neill: *Semi-Riemannian geometry*, Academic Press, New York, 1983

[O1] A.L. Onishchik: *Flag supermanifolds, their automorphisms and deformations*, in "The Sophus Lie Memorial Conference" (Oslo, 1992), 289-302, Scand. Univ. Press, 1994

[O2] A.L. Onishchik: *On the rigidity of super-Grassmannians*, Ann. Global Anal. Geom. **11** (1993), no. 4, 361-372

[O3] A.L. Onishchik: *Topology of transitive transformation groups*, Barth Verlagsgesellschaft, Leipzig, Berlin, Heidelberg, 1994

[O-V] A.L. Onishchik, E.B. Vinberg: *Lie groups and algebraic groups*, Springer, Berlin, Heidelberg, 1990

[P-P-S] H. Pedersen, Y.-S. Poon, A. Swann: *Hypercomplex structures associated to quaternionic manifolds*, Preprint Erwin Schrödinger Institut 426 (1997) (21 pp.)

[P] I.I. Piatetskii-Shapiro: *Geometry of classical domains and the theory of automorphic forms*, Gordon and Breach, 1969

[P-S] Y.S. Poon, S.M. Salamon: *Quaternionic Kähler 8-manifolds with positive scalar curvature*, J. Differential Geom. **33** (1991), no. 2, 363-378

[S1] S.M. Salamon: *Quaternionic Kähler manifolds*, Invent. Math. **67** (1982), 143-171

[S2] S.M. Salamon: *Riemannian geometry and holonomy groups*, Pitman Research Notes in Mathematics Series 201, Longman Scientific & Technical, Harlow Essex, 1989

[S3] S.M. Salamon: *On the cohomology of Kähler and hyper-Kähler manifolds*, Topology **35** (1996), no. 1, 137-155

[Sch] T. Schmitt: *Super differential geometry*, Akademie der Wissenschaften der DDR, Report R-MATH-05/84 (187 pp.)

[Schw] L.J. Schwachhöfer: *On the classification of holonomy representations*, Habilitationsschrift, Leipzig, 1998

[S-W] S. Shnider, R.O. Wells: *Supermanifolds, super twistor spaces and super Yang-Mills fields*, Presses de l'Université de Montréal, 1989

[St] A. Strominger: *Special geometry*, Commun. Math. Phys. **133** (1990), 163-180

[Sw1] A. Swann: *Hyper-Kähler and quaternionic Kähler geometry*, Math. Ann. **289** (1991), 421-450

[Sw2] A. Swann: *Homogeneous twistor spaces and nilpotent orbits*, preprint 97/17, Department of Mathematical Sciences, University of Bath, Bath, 1997

[T] T. Takahashi: *Sasakian manifold with pseudo-Riemannian metric*, Tôhoku Math. Journ. **21** (1969), 271-290

[V1] E.B. Vinberg: *Invariant linear connections in a homogeneous space*, Trudy Moskov. Mat. Obsc. **9** (1960), 191-210

[V2] E.B. Vinberg: *The theory of convex homogeneous cones*, Transactions of the Moscow Math. Soc. **12** (1963), 340-403

[V-G] E.B. Vinberg, S.G. Gindikin: *Kähler manifolds admitting a transitive solvable group of automorphisms*, Mat. Sb. (N.S.) **116** (1967), 357-377

[W1] J.A. Wolf: *Complex homogeneous contact manifolds and quaternionic symmetric spaces*, J. Math. Mech. **14** (1965), 1033-1047

[W2] J.A. Wolf: *The action of a real semisimple group on a complex flag manifold I. Orbit structure and holomorphic arc components*, Bull. Amer. Math. Soc. **75** (1969), 1121-1237

[W3] J.A. Wolf: *The Stein condition for cycle spaces of open orbits on complex flag manifolds*, Ann. of Math. (2) **136** (1992), no. 3, 541-555

[W4] J.A. Wolf: *Flag manifolds and representation theory*, in "Geometry and representation theory of real and p-adic groups" (Cordoba, 1995), 273-323, Progr. Math. **158**, Birkhäuser, Boston, 1998

[W5] J.A. Wolf: *Cayley transforms and orbit structure in complex flag manifolds*, Transform. Groups **2** (1997), no. 4, 391-405

[Wo] T. Wolter: *Homogene Mannigfaltigkeiten nichtpositiver Krümmung*, Dissertation, Zürich, 1989

[Z] B. Zumino: *Supersymmetry and Kähler manifolds*, Phys. Lett. **87B**, (1979), no. 3, 203-206

Editorial Information

To be published in the *Memoirs*, a paper must be correct, new, nontrivial, and significant. Further, it must be well written and of interest to a substantial number of mathematicians. Piecemeal results, such as an inconclusive step toward an unproved major theorem or a minor variation on a known result, are in general not acceptable for publication. Papers appearing in *Memoirs* are generally longer than those appearing in *Transactions*, which shares the same editorial committee.

As of May 31, 2000, the backlog for this journal was approximately 7 volumes. This estimate is the result of dividing the number of manuscripts for this journal in the Providence office that have not yet gone to the printer on the above date by the average number of monographs per volume over the previous twelve months, reduced by the number of volumes published in four months (the time necessary for preparing a volume for the printer). (There are 6 volumes per year, each containing at least 4 numbers.)

A Consent to Publish and Copyright Agreement is required before a paper will be published in the *Memoirs*. After a paper is accepted for publication, the Providence office will send a Consent to Publish and Copyright Agreement to all authors of the paper. By submitting a paper to the *Memoirs*, authors certify that the results have not been submitted to nor are they under consideration for publication by another journal, conference proceedings, or similar publication.

Information for Authors

Memoirs are printed from camera copy fully prepared by the author. This means that the finished book will look exactly like the copy submitted.

The paper must contain a *descriptive title* and an *abstract* that summarizes the article in language suitable for workers in the general field (algebra, analysis, etc.). The *descriptive title* should be short, but informative; useless or vague phrases such as "some remarks about" or "concerning" should be avoided. The *abstract* should be at least one complete sentence, and at most 300 words. Included with the footnotes to the paper should be the 2000 *Mathematics Subject Classification* representing the primary and secondary subjects of the article. The classifications are accessible from `www.ams.org/msc/`. The list of classifications is also available in print starting with the 1999 annual index of *Mathematical Reviews*. The Mathematics Subject Classification footnote may be followed by a list of *key words and phrases* describing the subject matter of the article and taken from it. Journal abbreviations used in bibliographies are listed in the latest *Mathematical Reviews* annual index. The series abbreviations are also accessible from `www.ams.org/publications/`. To help in preparing and verifying references, the AMS offers MR Lookup, a Reference Tool for Linking, at `www.ams.org/mrlookup/`. When the manuscript is submitted, authors should supply the editor with electronic addresses if available. These will be printed after the postal address at the end of the article.

Electronically prepared manuscripts. The AMS encourages electronically prepared manuscripts, with a strong preference for AMS-LaTeX. To this end, the Society has prepared AMS-LaTeX author packages for each AMS publication. Author packages include instructions for preparing electronic manuscripts, the *AMS Author Handbook*, samples, and a style file that generates the particular design specifications of that publication series. Though AMS-LaTeX is the highly preferred format of TeX, author packages are also available in AMS-TeX.

Authors may retrieve an author package from e-MATH starting from `www.ams.org/tex/` or via FTP to `ftp.ams.org` (login as `anonymous`, enter username as password, and type `cd pub/author-info`). The *AMS Author Handbook* and the *Instruction Manual* are available in PDF format following the author packages link from `www.ams.org/tex/`. The author package can be obtained free of charge by sending email to `pub@ams.org` (Internet) or from the Publication Division, American Mathematical Society, P.O. Box 6248, Providence, RI 02940-6248. When requesting an author package, please specify $\mathcal{A}\mathcal{M}\mathcal{S}$-LaTeX or $\mathcal{A}\mathcal{M}\mathcal{S}$-TeX, Macintosh or IBM (3.5) format, and the publication in which your paper will appear. Please be sure to include your complete mailing address.

Sending electronic files. After acceptance, the source file(s) should be sent to the Providence office (this includes any TeX source file, any graphics files, and the DVI or PostScript file).

Before sending the source file, be sure you have proofread your paper carefully. The files you send must be the EXACT files used to generate the proof copy that was accepted for publication. For all publications, authors are required to send a printed copy of their paper, which exactly matches the copy approved for publication, along with any graphics that will appear in the paper.

TeX files may be submitted by email, FTP, or on diskette. The DVI file(s) and PostScript files should be submitted only by FTP or on diskette unless they are encoded properly to submit through email. (DVI files are binary and PostScript files tend to be very large.)

Electronically prepared manuscripts can be sent via email to `pub-submit@ams.org` (Internet). The subject line of the message should include the publication code to identify it as a Memoir. TeX source files, DVI files, and PostScript files can be transferred over the Internet by FTP to the Internet node `e-math.ams.org` (130.44.1.100).

Electronic graphics. Comprehensive instructions on preparing graphics are available at `www.ams.org/jourhtml/graphics.html`. A few of the major requirements are given here.

Submit files for graphics as EPS (Encapsulated PostScript) files. This includes graphics originated via a graphics application as well as scanned photographs or other computer-generated images. If this is not possible, TIFF files are acceptable as long as they can be opened in Adobe Photoshop or Illustrator. No matter what method was used to produce the graphic, it is necessary to provide a paper copy to the AMS.

Authors using graphics packages for the creation of electronic art should also avoid the use of any lines thinner than 0.5 points in width. Many graphics packages allow the user to specify a "hairline" for a very thin line. Hairlines often look acceptable when proofed on a typical laser printer. However, when produced on a high-resolution laser imagesetter, hairlines become nearly invisible and will be lost entirely in the final printing process.

Screens should be set to values between 15% and 85%. Screens which fall outside of this range are too light or too dark to print correctly. Variations of screens within a graphic should be no less than 10%.

Inquiries. Any inquiries concerning a paper that has been accepted for publication should be sent directly to the Electronic Prepress Department, American Mathematical Society, P. O. Box 6248, Providence, RI 02940-6248.

Editors

Selected Titles in This Series

(Continued from the front of this publication)

669 **Seán Keel and James McKernan,** Rational curves on quasi-projective surfaces, 1999

668 **E. N. Dancer and P. Poláčik,** Realization of vector fields and dynamics of spatially homogeneous parabolic equations, 1999

667 **Ethan Akin,** Simplicial dynamical systems, 1999

666 **Mark Hovey and Neil P. Strickland,** Morava K-theories and localisation, 1999

665 **George Lawrence Ashline,** The defect relation of meromorphic maps on parabolic manifolds, 1999

664 **Xia Chen,** Limit theorems for functionals of ergodic Markov chains with general state space, 1999

663 **Ola Bratteli and Palle E. T. Jorgensen,** Iterated function systems and permutation representation of the Cuntz algebra, 1999

662 **B. H. Bowditch,** Treelike structures arising from continua and convergence groups, 1999

661 **J. P. C. Greenlees,** Rational S^1-equivariant stable homotopy theory, 1999

660 **Dale E. Alspach,** Tensor products and independent sums of $\mathcal{L}_p$-spaces, $1 < p < \infty$, 1999

659 **R. D. Nussbaum and S. M. Verduyn Lunel,** Generalizations of the Perron-Frobenius theorem for nonlinear maps, 1999

658 **Hasna Riahi,** Study of the critical points at infinity arising from the failure of the Palais-Smale condition for n-body type problems, 1999

657 **Richard F. Bass and Krzysztof Burdzy,** Cutting Brownian paths, 1999

656 **W. G. Bade, H. G. Dales, and Z. A. Lykova,** Algebraic and strong splittings of extensions of Banach algebras, 1999

655 **Yuval Z. Flicker,** Matching of orbital integrals on $GL(4)$ and $GSp(2)$, 1999

654 **Wancheng Sheng and Tong Zhang,** The Riemann problem for the transportation equations in gas dynamics, 1999

653 **L. C. Evans and W. Gangbo,** Differential equations methods for the Monge-Kantorovich mass transfer problem, 1999

652 **Arne Meurman and Mirko Primc,** Annihilating fields of standard modules of $\mathfrak{sl}(2,\mathbb{C})^\sim$ and combinatorial identities, 1999

651 **Lindsay N. Childs, Cornelius Greither, David J. Moss, Jim Sauerberg, and Karl Zimmermann,** Hopf algebras, polynomial formal groups, and Raynaud orders, 1998

650 **Ian M. Musson and Michel Van den Bergh,** Invariants under Tori of rings of differential operators and related topics, 1998

649 **Bernd Stellmacher and Franz Georg Timmesfeld,** Rank 3 amalgams, 1998

648 **Raúl E. Curto and Lawrence A. Fialkow,** Flat extensions of positive moment matrices: Recursively generated relations, 1998

647 **Wenxian Shen and Yingfei Yi,** Almost automorphic and almost periodic dynamics in skew-product semiflows, 1998

646 **Russell Johnson and Mahesh Nerurkar,** Controllability, stabilization, and the regulator problem for random differential systems, 1998

645 **Peter W. Bates, Kening Lu, and Chongchun Zeng,** Existence and persistence of invariant manifolds for semiflows in Banach space, 1998

644 **Michael David Weiner,** Bosonic construction of vertex operator para-algebras from symplectic affine Kac-Moody algebras, 1998

643 **Józef Dodziuk and Jay Jorgenson,** Spectral asymptotics on degenerating hyperbolic 3-manifolds, 1998

642 **Chu Wenchang,** Basic almost-poised hypergeometric series, 1998

For a complete list of titles in this series, visit the AMS Bookstore at **www.ams.org/bookstore/**.

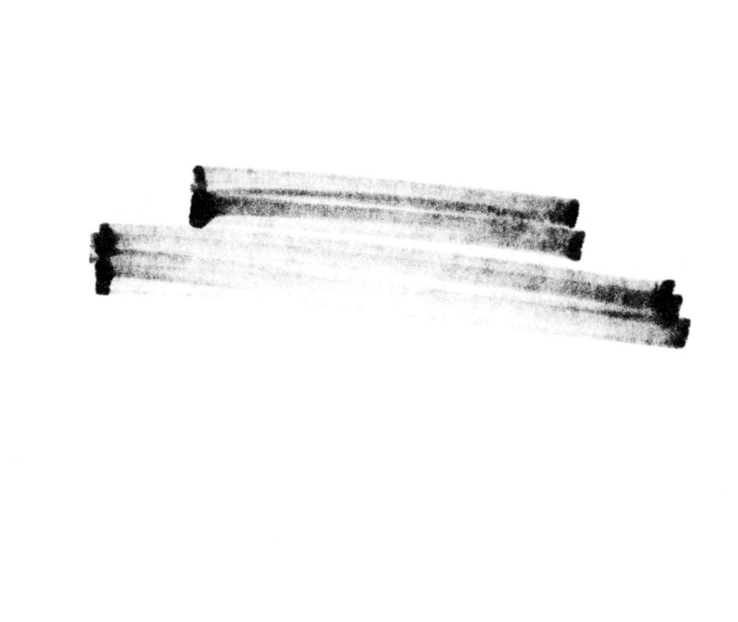

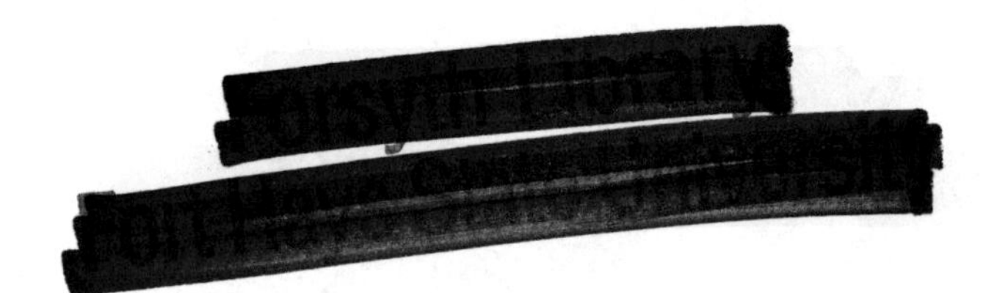